THE FATES OF NATIONS

A BIOLOGICAL THEORY
OF HISTORY

Paul Colinvaux

SIMON AND SCHUSTER
NEW YORK

COPYRIGHT © 1980 BY PAUL COLINVAUX
ALL RIGHTS RESERVED
INCLUDING THE RIGHT OF REPRODUCTION
IN WHOLE OR IN PART IN ANY FORM
PUBLISHED BY SIMON AND SCHUSTER
A DIVISION OF GULF & WESTERN CORPORATION
SIMON & SCHUSTER BUILDING
ROCKEFELLER CENTER
1230 AVENUE OF THE AMERICAS
NEW YORK, NEW YORK 10020

SIMON AND SCHUSTER AND COLOPHON ARE TRADEMARKS OF
SIMON & SCHUSTER

DESIGNED BY EVE METZ
MANUFACTURED IN THE UNITED STATES OF AMERICA

1 2 3 4 5 6 7 8 9 10

LIBRARY OF CONGRESS CATALOGING IN PUBLICATION DATA

COLINVAUX, PAUL A.   DATE.
    THE FATES OF NATIONS

    BIBLIOGRAPHY: P.
    INCLUDES INDEX.
    1. HISTORY—PHILOSOPHY.   2. MILITARY HISTORY.
3. HUMAN ECOLOGY.   4. NICHE (ECOLOGY)   I. TITLE.
D16.9.C597   901   80-12735
ISBN 0-671-25204-6

To the John Simon Guggenheim Memorial Foundation who helped me to time off from researching the ancient climates of Bering Strait and the Galapagos Islands, letting me indulge my yearning to read history and politics for a year, in a little mews house close to Hyde Park.

PAUL COLINVAUX

Columbus
16 April 1980

Contents

CHAPTER ONE

THE FATES
OF NATIONS

HISTORY IS about the doings and desires of unusual animals, members of the species *sapiens* in the genus *Homo*. In the last few thousand years this strange species has so increased its population that it has almost overrun the earth, and it has changed its styles of living from humble hunter to civilized human. These changes in habit and number are the cause of cycles in history and they settle the fates of individual nations.

The actual record of human affairs is a tale of battles and rebellion, conquest and trade, colonies, tyranny, liberation and empire. The nations come and go fighting, and the focus of power shifts from one nation to another. History repeats itself as an emergent people expands against its neighbors, lives on in splendor and freedom for a time, and then slides away into ignominious eclipse. For us of the West the names of ancient empires of the occident haunt us still—Egypt, Persia,

Greece, Rome all climbed to power on battle, but all lasted to see their power slide away until foreign soldiers swarmed over the lands that once they held. More recently, in the forging of modern Europe, empires were torn in turn from Portugal, Spain, Sweden, France, the Netherlands and Germany, and at last the sun set, amidst the sound of guns, on the British Empire, the largest portfolio of real estate ever put together in the history of the world.

That changes in habit and, above all, in human numbers ride alongside the cycles of history seems apparent enough. Consider how often a great nation or empire is built by people who once were few and unimportant— the English, the Romans, the American colonists, the followers of Muhammad; for each of these, the climb in numbers of their descendants as they triumphed is equaled only by the massive changes in their ways of life. But changes in number do not come from victory in battle; they come from success at breeding. And changing habits modify ancient ways which natural selection gave us when it cut us off from the hominid line tens of thousands of years ago.

To understand history we must first know how people breed and how they change their ways, which are jobs for an ecologist. Ecological research is usually designed to answer questions such as, "What limits the population of this animal?" or, "Why are these animals doing what they do?" In the last fifteen years ecologists have gone far with general answers to questions like these and we think we understand the limits to number and habit in most species other than our own. And yet people are animals too, fashioned by natural selection like other animals. The very fates of nations must reflect the populations and habits of their people and ought, therefore, to be understood by ecological analysis.

Ecologists hold a Darwinian view of life. Only animals fit both to get their food and to breed most excellently well can leave a long posterity. We expect everything an animal does to be well suited to getting food, for escaping death, or for raising young; above all for raising young. Natural selection works to make all animals into factories for producing fresh animals out of raw materials wrested from the environment. Ecologists joke to one another that they are the most ordinary of people because they spend their time thinking of either sex or food. The jibe is true in that most theoretical ecology turns on how animals and plants get food and reproduce. The peoples of whole nation-states get food and reproduce in a great cooperative enterprise. The history of nations, therefore, is a fit subject for ecology.

Once upon a time humans lived according to the same ground rules as all other animals, but that was at the dawn of history, in the days of the last ice age, fifty thousand years ago when we first emerged as *Homo sapiens*. Since then we have changed some of the ground rules, but not all of them. In particular, a close examination of how we conduct our family life will show that we have made no significant change in our habits of reproduction from that remote ice-age past. We breed like animals still.

But we have changed the rules that once fixed our way of life. Other animals have very little choice about how they must live and ancient people were much the same; they gathered what food could be found, crawled into a shelter, and made do. History really began when we began to think how we might live better and so broke the animal rules. For nine thousand years now, we have experimented constantly with new ways of getting a living, inventing farming, cities, industry and government, then endlessly improving each. The achievements of nation states are the records of these experiments. But other, unchanging rules have been with us all the time too. In particular, our animal breeding habits have pressed ever larger populations to live in each succeeding society. History has been a long progression of changing ways of life and changing population, the one always chasing the other. War, trade and empire are the results.

I find that ecological analysis gives the causes of the great aggressive wars, and the ways in which they are fought. Much of this book, therefore, is about battle and weapons, for, without understanding what really happened in the great wars, and why the victor won, we cannot understand the causes. I do not apologize for turning ecology to the study of war, because only by knowing why the great attacks are made have we any hope of bringing the sequence of aggressions to an end. This end is not yet near, but the ecology of battles can tell us what the next, and nuclear, events will be.

Many of the well-known scourges of our kind, including poverty and social oppression, turn out to be clearly predicted by ecological theory, poverty in particular being an inevitable consequence of changing habits and of reproductive drive. Links between family size, wealth and liberty are also understood when subjected to ecological analysis. Even freedom itself can be defined in ecological terms, and an ecologist's view of history points with accuracy to the nations which should be able to retain freedom.

And yet there is nothing in my analysis of the sorts of thing which "ecology" has come to imply in public debate. There is nothing of catastrophe, of doom, of people helpless before the forces of a malignant or abused environment. Nor are there fairy tales of technological plenty to come or social views of history which divorce people from their animal nature and the physical world in which they live. I believe all these schools of modern thought to traffic in nonsense, for I have reviewed most of the ecological fables and have found them wanting.

My thesis is thus that the fates of whole peoples may be understood, and even predicted in outline, from a knowledge of the ecological nature of our kind. I seek to test this ecological theory of history against some of the more familiar and mighty episodes of our past: the rise and fall of the Mediterranean civilizations; the savage upheavals of warrior peoples from the Asiatic steppe; the American experiment with liberty; and the rise of the Western power, with its unprecedented weapons and its knack for ever-accelerating change.

The fates of individual nations have often been seen to hinge on chance: on the career of a remarkable man, on a code of morality perfectly suited to the times; on the happenstance of battle. It would be a dried-out person who wished to deny the roles these things have played as arbiters of our fate. But chance acts only on the stage set for it by the fundamental ecological nature of the human kind. Great leaders, fine moralities, or superior weapons of war are important only as they act on populations and ways of life that then exist. It is these populations and habits which we can understand from ecological principles. And from an understanding of these principles it becomes possible to look into the future, tremulously perhaps, but with some conviction of the sorts of things which are likely to happen in the next few centuries.

CHAPTER TWO

ALEXANDER AT ARBELA

THE GREATER climactics of history tend to be marked by desperate wars as the old order gives way to the new at a divide set by battle. We remember these wars, often enough, by the names of the great captains who fought in them: Hannibal, Caesar, Napoleon. The story of the fighting that gave Alexander the name of "Great" shows the kind of thing that happens, and the aggression that Alexander wrought some two thousand five hundred years ago describes many of those things that an ecological hypothesis must predict and explain.

Alexander fought his most celebrated fight at a little place called Arbela, in what is now part of Turkey. It was there that the army of an Asian empire, under its Persian king, had been drawn up to meet him, an enormous force; prepared, well supplied, well led, brave, set out in proud national regiments on ground of its own choosing. It was spread in a long line across a sandy plain, a wall of men and horses. On the wings was splendid cavalry from nomadic peoples whose usual days were spent on horseback and who had the best weapons and armor that money could buy. In the center was stubborn professional infantry under Darius, their king. And stretching far and wide in front of them was a cleared and leveled tract of land, a parade ground for the dash of cavalry and the sweep of chariots. All that competence and experience could suggest to improve the chances of battle had been done. The defenders of the Persian Empire waited in their prepared position the assault of the expedition from Greece.

And Alexander came against that huge Asian army early one morning with forty-seven thousand men, perhaps a tenth of the number waiting for him in their long ranks behind the cleared and open space. Once committed to the open, Alexander's little army would be enveloped in clouds of dashing squadrons, and attacked by horsemen and missile men from all sides in a living sea of war in which his small force could be nibbled away and destroyed. This was the Persian plan. But the little Greek army came on, in two divisions with columns at the sides, ready to form a hollow square against which the sea could break. And it marched with the grim and rhythmic crunch of highly trained men, clanking a little where arms and armor touched, with snuffles and whinnies from the horses on the flanks but with seldom a sound from a human voice. And its color was the color of armor plate.

The Greek and Macedonian soldiers of Alexander wore helmets, breastplates, steel-ribbed skirts; and greaves to protect their lower legs. Each man had a solid, round shield, bossed with metal, and they huddled together in echelon so that their shields overlapped. There was little to be seen of a man from in front except armor. The whole array bore down on an enemy line as the cold, gray edge of a slab of steel; massive, ponderous, dreadful. The men of each division were arranged in sixteen ranks, carefully spaced, each man placed so that he might cover the gaps between his fellows, his helmet, breastplate and shield a portion of the communal armor plate. And each man held a spear sixteen or twenty feet long, a weapon for remote killing with a brutal iron point.

The Greek soldiers were drilled to turn in any direction and to reach out between as many as five files of their comrades to plunge that iron point into the bellies of approaching enemies. Their sixteen ranks of spears and armor formed the celebrated Macedonian Phalanx, the invention of Alexander's father, Philip, and the most impersonal agent for killing brave adversaries which the ancient world had discovered. This phalanx crunched deliberately down on the young men of Asia as they stood in their thousands ready to defend their homes.

Darius of Persia had pondered the problem of this awful phalanx. Somehow it must be broken into so that his seas of soldiers could burst inside. Darius hoped to hold it on the flat land which his engineers had cleared, then to plunge holes in it with a fleet of war chariots. But it moved down on him crabwise, sliding across his front as it closed. Soon it would be at the end of his line, clear of the ground prepared for chariots, no longer in a position to be enveloped in that sea of war.

Darius ordered his cavalry to stop the phalanx and the sea of war was made; the ancestors of Sihk and Turk hurled themselves on the aggressor. They fared well against the Greek horsemen, being man for man, and perhaps slightly better armed. They peeled the cavalry shield from the ends of the phalanx, but against the phalanx itself the horsemen could do nothing. To charge it was to have horse and man impaled.

So Darius kept his front clear and sent in his fleet of chariots. They swept down with a rumbling roar, a hundred flying plumes of dust, a hundred pointed rams jutting forward from the shafts between the horses. But running in front of the phalanx were lightly armed Greeks who dodged before the chariots, threw javelins, and shot arrows into them as they went by. For many Persian drivers the combination of torment by missile and the sight of that squat gray slab ahead was too much, and they turned before they closed. A few, lucky and resolute, rattled right to the very line of spears. But the spears rose, and the files stepped aside, precisely, as they had done to the calls of their sergeants in practice. Clear lanes were left through the phalanx down which the horses plunged to draw the chariots on: through, and past the Greeks waiting in the rear to stab the drivers in the back. The chariots were spent. And there was still no way for the sea of war to burst inside the ranks of Greeks.

Only butchery was to come. The phalanx stamped down into the Persian line, the spears stabbing, impaling, killing from comparative safety, almost as men have killed from the safety of machines in later times. Darius' heart failed him as the butchering slab of steel trampled ever closer down the line, and he turned his chariot and fled.

When the killing was over, Alexander only had to march on across Asia, dealing with lesser armies in the same way, giving the government to his own people, settling in other people's lands the Greeks who streamed after him, giving his arrogant name to two dozen cities. People, marveling at the extent of the conquest, ascribed it all to personal prowess and called Alexander "the Great."

The victory at Arbela and the subsequent imposition of the Greek way of life onto all the lands and peoples known to Greek geography was one of the great climactic events of history. An expanding people had engulfed another and a new code of civilization and culture was recognized as the norm, remaining the well-spring of human aspiration for centuries. The grandeur of the achievement seems to justify Alexander the Great's title, and we are apt to forgive him that he had to wade through slaughter to his throne. We are also apt to forget the part of technology and a better social system in his triumph. For what really made his victories so complete was the superior technique, equipment and discipline of his army. If the Macedonian phalanx was used properly, no Asiatic troops of the day could stand against it. At Arbela, Darius had done all that a competent general could. That he was not a captain in the same rank as Alexander is beside the point. In truth he had no chance. His chariots had no more hope of breaking a Macedonian phalanx than had the Polish cavalry of breaking the German panzer divisions in 1939. Superior technique won the day.

Alexander had inherited his splendid army, with its deadly phalanx, from his father, and from his father's generation. Because he was a clever young man with the wit to use what was given him we must not forget that others forged his instrument. Nor must we forget that it was the instrument that won the war. The Greeks and Macedonians who followed Alexander to victory did not conquer because they were braver than the people they thrust down. All people are brave. Nor did they win primarily because they had the better general, though they did have that. They won because they had better weapons and technique.

A critical examination of all the massive, successful aggressions of history will show this same general rule: the victors are the side with the better weapons or the more advanced technique. It was brooding on the fate of Darius and his army at Arbela that first made this fact so vivid to me. What could one do when faced with that frightful armored phalanx, bristling like a hedge-hog with spears? The answer is, "Nothing." You could not train a phalanx of your own, because it took years and the tradition of a culture to produce so disciplined a thing. You could perhaps try to buy yourself a phal-anx, to enlist mercenaries who came from a culture trained to that extreme discipline of war. Darius did his best here too, recruiting a contingent of renegade Greeks with all their armor. They were soldiers who had no comfortable life in Greece, knowing only the trade of war, outcast men hating Alexander and the kind of Greece he stood for. These renegades fought for Darius to the last man. Killing them was the bloodi-est business that Alexander had to do at Arbela. It was only where the Greek mercenaries stood that the Persian line held for any significant time. But there were not many Greek infantry to be bought, and Persians could learn neither the weapons nor the skill in line.

When Darius's mercenary Greek soldiers were all dead, there was nothing for a Persian to do but surren-der or be killed. This is the same stark choice that has been faced by all defenders of states about to be subjugated by the conquering armies that build empires. They have no chance.

Conquering techniques change with time. Romans were to find the technical answer to the phalanx and call it the "legion." Other winning tricks were to be the disciplined horse archers of Genghis Khan; Spanish "battles" of infantry with their massed pikes and corners of muskets; the mobile light cannon of Gustavus Adolphus; and massed tanks closely supported by dive-bombers. But all have in common that they give victory to the people who invent them. Since the fates of nations often seem set by conquests made possible by superior technique, I shall describe the courses of many decisive battles in this book. Yet it is the reason for the invention of conquering techniques that must be found.

Ancient Greece had a civilization so fine that we still study it for profit. Yet, under Alexander, these civilized folk set about armed robbery on a national scale. And because the robbery was successful, and other people's lands were incorporated into the real estate controlled by Greeks, the man who led the exploit is called "the Great."

If we would understand the achievements of Alexander, and indeed of the men of similar climactic events of history, we must know what led their people to invent new techniques of war and what gave them the will for aggression. The answers can be found in the histories of the aggressor states. Always we will find that the great conquest was preceded by decades of strife at home, by signs of a turbulent expansionist society, by adventurous commerce. This was surely so for Greece, whose fighting and inventiveness we still study. The Greeks steadily invented a better and better life, a more civilized life, a life richer than that of neighboring states.

They also quarreled among themselves as their expansionist ideas took to including their countrymen and their countrymen's lands into the orbits of their ambitions. Skills in living and better technology inevitably came to be applied to war as well as to peace.

In such a turbulent expanding society it must only be a matter of time before some general cleverer than the rest succeeds in gathering the national armies to his personal allegiance and loosing what has become an irresistible force against adjacent states. These have neither the military means nor the social order to withstand him. The seal of success is set on the strivings and inventions of several generations, by the establishment of an empire, and the general who performs the climactic act takes his place in the list of Great Captains.

In the long view, Alexander was no more than the instrument of a destiny already made certain by the evolution of Greek society. Ending of local wars had eventually united Greece. Alexander's father, Philip of Macedon, had bequeathed him his incomparable army, and the finest teachers of his age, including Aristotle, had seen to his education. Alexander's genius let him see what all this had made possible, but, if he had not been worthy of the task, history would have found another before long. The great captain was but the instrument of a social process. To understand the process and the aggressions to which it led we must look to social mechanisms.

Greeks went to war for plunder, for living space, for land for the younger sons and the underprivileged. They obviously thought they needed these things. For a couple of centuries Greeks had been improving their civilization, making it possible for people to aspire to more ample lives. Technical ingenuity let Greeks provide the goods and services, and the food, which met the physical requirements of those ample lives. But a constant aspiration to live better must mean a constant demand for more goods and services.

When we talk about providing goods and services we approach subjects on which an ecologist can comment from the knowledge of the ecological profession. Ecologists study animal numbers and animal distributions as the results of silent compacts between individuals and species to share out resources. We talk of animal "niches," their individual places in the natural order, as being bounded by sets of resources. Niche theory explains both the spread of animals in space and the numbers in their populations. In ecological language, the expansion of Greek society involves changes in the niches of the Greek people.

But ecologists will tell you that the supply of resources also determines breeding success in animals. Give an animal population a larger resource base, and the animals will multiply more rapidly until all the new resources are rationed out. It must at least be possible that the human animal can respond to more resources not only by better living but also by providing more people.

Ancient peoples left us no counts of their numbers. Their governments left few printed censuses. But their written exploits are often eloquent enough of changing number. When the course of history runs from scattered herdsmen and shifting farmers to the dense populations of cities, a crowding of people seems beyond dispute. Once the cities are built and thriving, the doings of the people, as in the Greek states, tell of mounting numbers. Greek cities founded colonies, true colonies, little overflow bits of Greece to which it was possible to export surplus people. The city states built

fleets to trade with other countries, a proceeding that
not only brought resources home to the parent city but
which also gave jobs for the traders in other people's
lands. These things strongly suggest rising numbers as
well as rising aspirations.

Certain it is that the people-exporting, trading, col-
ony-building Greek city-states crossed one another and
went to war. They fought among themselves in the century when Greek fought Greek. Then, finally, they
fused together in a swelling national overflow behind
their captain Alexander, and conquered the Persian
Asia of Darius.

It is probably true that an expanding population with
a rising expectation of life lay behind the empire of
Alexander. The very technical ingenuity which made
possible the rising expectations also gave Greeks the
military instrument to take resources from their neighbors. They needed the land, they had the power to take
it, so they took it. This is the true explanation of Greek
aggression and Alexander's conquests.

Behind all the great climactic struggles of history we
will find symptoms of an expanding population. Whenever people have been ingenious so that the quality of
their lives has improved they have let their numbers
rise. The demand for more resources for the better life
has always been more than the prevailing political systems could provide. And the grand themes of history
have been the result: repressions, revolutions, libera-
tions and always, in the end, aggressive war.

Perhaps little wars and petty repressions can often be
explained as being caused by no more than human
wickedness and animal passions, as various social and
biological writers have argued. But all the truly great
wars of history, those that ended with shifts of peoples

and the remaking of maps, were caused by increases in the numbers of people and associated increases in demand. We can examine the wars, the growths, and the falls of civilizations from ecological principles which describe how resources must be divided between people and which show consequences of changing the numbers of those people. From this study a predictive theory for the fates of civilizations, including our own, will emerge.

CHAPTER THREE

THE HUMAN NICHE

P EOPLE, LIKE other animals, have niches in life.
Much of the more violent sort of history comes about as
individual people struggle for what they perceive to be
their proper ecological niche. And people also have a
breeding strategy, without which they would have no
descendants. It is the excellence of our breeding strategy that lets us provide the crowds who staff the industries and armies of a great civilization.

By "niche" an ecologist means all the things about a
kind of animal that let it live: its way of feeding, what it
does to avoid enemies, how it is fitted to the place in
which it must dwell.

By "breeding strategy" we mean all the special things
the species does to turn resources into babies, whether
it lays eggs or bears its young alive, whether it breeds
only in the spring, or every year, or only once in a lifetime. Because of its special niche an animal can survive

and win food. Through its breeding strategy it turns much of this food into babies. When people work to win resources to raise a family they are following in a refined way the animal proceeding of living in a niche, and then turning some of the resources they get into their descendants.

The word "niche" seems to have been borrowed by ecologists from church architecture, where it means a slot or place in a wall where a statuette may stand. To an architect the word "niche" means a physical place, a fixed location in a pile of bricks. But an ecologist uses the word "niche" to mean a place in the web of life. The niche of an animal reflects its role in a community: to eat its prey, to be eaten by its hunters, to occupy a place in a habitat. So when an ecologist talks of "niche," it is the animal's job that is meant, rather than where the animal can be found.

Every species has its niche, and the niche of each species is different from that of all other species. This fundamental maxim of ecology perhaps becomes clearer when we substitute the word "profession" for "niche." Every species has a unique profession to which all the individuals of that species are trained. The jobs available in each animal profession set the size of the animal population.

Consider the niche, or profession, of the familiar gray squirrel of a woodlot or garden. Gray squirrels get much of their living from seeds, both young seeds forming on branches in summer and fallen seeds scattered through the litter of the woodland floor in autumn and winter. These seeds, supplemented by birds' eggs, some insects and assorted greenery, are the squirrel's resource. Living in the squirrel niche involves tapping this resource in an effective and specialized way. Each individual must run and climb with great agility; it must also be able to dig in the right places; and it must have the proper habit of burying little caches of nuts in the fall where its winter searches will come on them again. These very special skills of the gray squirrel are all parts of its niche. Together they make a package of food-hunting habits and abilities which carry a squirrel through the year. They are very essential skills to the squirrel; but there is more to the niche of any species than just being good at nosing out food, because you must stay alive to enjoy your food.

The successful squirrel must be as good at looking out for cats, dogs, mink, weasels, owls, hawks and the like as it is at looking out for nuts. Each squirrel must hark to the chatter of other squirrels. It must be ever ready with that quick dash to a friendly tree. It must also be able to build a nest (a "dray" the old naturalists called it) of leaves in the fork of a tree, to arrange to have young there, to retreat to this shelter and huddle against other squirrels on a cold day, to conserve energy in the dray during bitter cold by lowering the body temperature and the pulse rate. And it must respect all the social customs and norms without which a squirrel cannot find a mate or leave offspring. All these things are part of the gray squirrel's niche, and every gray squirrel must be good at doing all these things right. Being a gray squirrel is a highly professional undertaking; amateurs would not survive at it.

There are other kinds of squirrels, mostly in shades of red rather than gray. All of them do some of the things that the gray squirrel does, particularly in the eating of nuts. Most live in trees, and are elegantly balanced like the gray with a long fluffy rudder of a tail. But each kind specializes on nuts in different ways.

Some specialize on seeds in pine cones and are programmed to do the things needed to get through the
year in a pine forest. Others live in different kinds of
broadleaf forest, or on the ground but all exploit seeds
and fruits typical of the places where they live.

Each of these kinds of squirrel has a specialized niche,
a different profession from all the rest. Each kind is
highly qualified for the trade it must follow in life, nicely
programmed to do all the right things so that it can live
on to exploit a particular pattern of food items, in a
particular place, and under a particular climate. But
there is very little give in any one squirrel program; it
cannot do what the other kinds of squirrel can do, or at
least not very well. This brings us to the most essential
quality of the niche of any animal: it is fixed.

A squirrel cannot change its way of life. It knows but
one trade. Squirrels, therefore, live only at the times
and places suited to the squirrel way of life, to the squirrel niche. And a result of most enormous importance
follows from this. The numbers of any kind of squirrel
that may live are fixed.

Niche sets the size of populations of all animal species.
This can be made particularly clear by following the
analogy of niche with a human profession. We have
talked of the profession of being a squirrel and of how
good at this profession a squirrel must be. Think instead
of the human profession of aeronautical engineering.
We know that the number of people who can earn their
living as aeronautical engineers is set by the job market
for these highly specialized skills. The number of people actually filling the niche of aeronautical engineering
cannot be altered by training more engineers in college
but only by making the aircraft industry boom. An ecol-

28

ogist would say that niche-space determines the population of the species "aeronautical engineer," just as niche-space determines the number of squirrels.

Similar arguments apply to all human professions, just as they apply to all kinds of animal niche. If, in any city, we turn out more lawyers from the local law school than there are jobs for lawyers, the surplus will not be able to follow the profession for which they have been trained. Every community has just so much room for lawyers, just so many niche-spaces in the law. The experiment of making an oversupply of lawyers has often been tried; and the consequence, frustrated people looking for a life as congenial as the one from which they were forced, is one of the small pressures that drive history.

Western societies have recently tried a large-scale experiment in flooding niche-space when they expanded the university population, particularly the graduate schools. A glut of Ph.D.s were trained for doctorates and wanted to be university professors. They were equipped for "professing" as a gray squirrel is equipped for squirreling. Universities have produced very large numbers of these presumptive professors, rather as if the squirrels had a very good year for raising young. But the number of professorships sets the opportunities for professing, and, by the late 1970s it became apparent that these professorial niche-spaces were all filled. Now surplus bearers of doctorates cannot accept the scholar's tenure, however *cum laude* their degrees, and must take to honest work for a living.

Scholars changing their jobs illustrate what an ecologist must look on as the most special thing about people: they can change. People have the quality, not shared by any other animal, of changing their niches. Surplus squirrels always die, but surplus scholars, lawyers and aeronautical engineers take up other trades. Yet it must be remembered that all human professions have this in common with animal niches, that the number of individuals following each profession, or niche, is absolutely set by the conditions of their ways of life. Niche sets number.

Squirrels of all kinds are numerous animals, very common where they are found, very abundant. It must follow that typical squirrel niche-spaces are correspondingly abundant and it is easy to see that this is so. Nuts and seeds grow almost everywhere, from the arctic to the desert, with particular concentrations in the forest. Squirreling is a sophisticated way at getting at this abundant resource, a profession that eats seeds, stores seeds, turns to other foods in lean times, and even sleeps away the worst times of all. So the various squirrel niches, each occupied by a distinct species, make nearly all squirrels common animals.

But other animals have niches that make them much less common than squirrels. Grizzly bears are a good example, because there is much in the bear's way of life which was once our own. Bears have a catholic diet, eating vegetables, seeds, berries, tubers and meat, just as we like to do. But grizzly bears are rather thin on the ground, even when not persecuted by rifles. You may see a bear now north of the Brooks Range in Alaska, in the open tundra country along the pipeline road. In a day's drive you have a good chance of seeing a single bear, moving across the tundra with the loose, bear gait, covering distance in its perpetual wandering. It is thought that a grown Alaska grizzly wanders thirty miles each day, but you will probably have to travel many more miles than thirty to be sure of seeing a wild bear, for they wander alone over great expanses.

Grizzly bears are rarish animals because the food resources of the bears are thinly spread and uncertain. If an animal would go a'bearing on the Alaskan tundra it has to be rare. It would have been easier to apply the grizzly bear's professional skills to the western plains, in the years before the European settlement. The Great Plains were more productive of fruit and meat for bears to catch, but the opportunities would still have been very restricted compared with the opportunities for being a squirrel. The bear niche would have set the grizzly bear numbers at a fairly low level, even on the presettlement plains.

But bears can reproduce rapidly. A typical bear family has several cubs in it, and a female bear can reproduce again and again. The young bears start life safely, because few animals are prepared to trifle with a baby belonging to a female grizzly. Yet the bears stay comparatively rare. The reason for the rarity is that there are few niche-spaces for bears. No degree of reproductive drive can force the environment to support more bears, and the admirable motherhood of a female grizzly cannot, and does not, make any difference to the numbers of bears that live.

In a world where the job opportunities of niche-space determine all populations it is obvious that reproduction never will make any difference to the numbers living. However well an animal breeds, whether it lays a dozen eggs or a thousand, this will make no difference to the size of the population. This is so strange a thing to say that it is well to repeat it before going on to describe breeding strategies, our own included. *The reproductive effort makes no difference to the eventual size of an animal population.*

In spite of the hopelessness of adding to the numbers of one's kind, every living thing perpetually breeds as if in a race for baby bonuses. Even an elephant couple can replace itself many times over, and salmon or mosquitoes have egg factories that look to have been designed by Henry Ford. Why all this terrible effort at breeding, if the population remains exactly the same? The answer is that every individual animal is driven on to breed by the anarchy of natural selection which pits neighbor against neighbor in a reproductive race.

Although a total population is set by niche, there is nothing in this to say whose offspring will make up that population. All the niche-spaces or jobs made vacant in the next generation by death are going to be filled by somebody's babies. This is where the rather stark proceeding called "natural selection" operates. The only family lines that survive to a long posterity are those which are successful in placing offspring in jobs. And the more jobs won by the family, the more certain the future. All individuals of all animal kinds are programmed to make and raise the largest possible number of surviving offspring. If they do not, they will be denied a posterity by the more busy breeding efforts of those around them. It becomes a mathematical certainty that the large healthy family will replace the small healthy family in evolutionary time.

One expectable and obvious consequence of this competition to make babies is a breeding strategy of mass production, which involves making a large and indefinite number of tiny eggs. When a mother is a mere egg factory, the more food she can get, the more eggs she lays. Female salmon, spiders, frogs, dandelions, mushrooms and butterflies all work as egg factories fitted to particular niches. It is by far the commonest breeding strategy. But it pays the price of causing very great waste. Not only are the numbers of hatched young ruinously in excess of the niche opportunities, but also each offspring is tiny. Its smallness means that it is almost certain to die soon, probably, in fact, by being eaten.

I call this breeding strategy of going for large numbers of tiny family starts "The small-egg gambit." The strategy makes the sort of crazy, anarchic sense characteristic of much that is fashioned through evolution by natural selection. Each parent is forced to waste resources on more and more tiny offspring simply because a neighbor is doing the same thing. If the females in the Jones family of salmon habitually lay ten eggs fewer than the females of the Smith family, then there will come a time when only Smiths are left. And so the ruinous game of desperate reproduction goes on, even though it cannot make any difference to the size of the population.

But people do not lay masses of tiny eggs. They hatch one egg at a time and carefully nurture it, at first inside the mother and then in a close family group, sometimes for as long as twenty years. Though this seems a conservative or abstemious approach to breeding, it is nothing of the kind. The breeding strategy of caring for a few young also meets natural selection's requirement that breeding be to the uttermost, and even excels in effectiveness the strategy of making eggs by the thousand. I call this breeding strategy to which we subscribe the "large-young gambit."

Raising a few large young, and looking after them, gives the advantage in that it avoids the wastage of both energy and food which follows from making young so helpless that most of them die. When an insect or a fish throws out masses of tiny eggs, the inevitable result is that most of the eggs or hatchlings are eaten by other animals so that the mother is busy feeding in her niche

33

to support a bunch of free-loading predators. Some of her young do survive, but the number that live may represent a poor return for her investment. She is like a casino gambler who places chips of low value on almost every number on the board, a procedure generally thought of as poor investment practice. But raising a few large young, as people do, avoids all this. If the thing is well done, so that the parents can support the chosen family of a few healthy infants, then there may be very few losses to predators or by chance death. Every bit of food and effort invested in making and supporting a baby will yield a return in a vigorous descendant able to compete for one of the niche-spaces in the next generation. Waste is kept to a minimum. The large young gambit may thus be thought of as the banker's approach to investing in reproduction, whereas mass-producing tiny eggs is a gambler's expedient. Bankers are known to be richer than gamblers.

Having a few large young, and looking after them, is the best way to press your descendants into the populations of the future. Very many abundant animals operate in this way, including all birds, viviparous reptiles and sharks, and all mammals. We are just one species among very many who conserve resources and bank them in young that stand a very good chance of succeeding us. Our breeding is cost-effective. And it is vital to master the point that our small families in no sense represent a modest breeding effort or reproductive restraint. On the contrary, we have been equipped with breeding habits as effective as any that are known for increasing the numbers of each couple's survivors.

But even for those animals with the prudent banking habits of the large-young gambit, there must still be a pressure to raise the greatest possible number of babies, producing a tendency to make the modest family hold just one youngster more. The tendency is blocked or balanced by the danger that lies in trying for too large a family. If a couple tries for one youngster too many there may not be enough food to go round and the whole brood may be in jeopardy. One youngster too *few*, and your neighbors' descendants will swamp yours. One youngster too *many*, and you tend to lose whole broods, so your neighbors again triumph. The family line that wins will be the one which starts with exactly the right family size: the largest number of babies that can be reared on the food available, and not one baby more. There will, therefore, be an optimum family size for any species using the large-young gambit, and habits which result in this optimum family will be preserved by natural selection.

Natural selection was undoubtedly alive and well when the breeding strategy of ice-age *Homo sapiens* was fashioned. Our ancestors were, therefore, adapted to make the best possible choice for the number of children in the family, given that "best" means the largest number for which there is a high probability of rearing them to maturity. So much seems certain. But in making that "choice" of the optimum family the primeval people had to choose the number they could support for ten to twenty years. Human parents could only add to the family every twelve months or so, which meant that choosing the "right" number of children was bound to be drawn-out business. Human sex and family life must be suited to coping with the special problems of humankind—only one child a year and many years of looking-after to follow.

People have strange sex habits. The females ovulate

roughly every lunar month and the males are apparently ready for sexual adventure every day of the year. This pattern is not quite without precedent in the animal kingdom, since a few monkeys come close to it, but it is a pattern that is extremely rare. It is far more usual for sex to be programmed with the seasons, or, more particularly, with some glut of food that can be used for feeding young. This is why birds, foxes and sheep give birth in the spring, even if it means mating in the depths of winter, as it does for foxes and sheep. But people ignore seasons altogether and go for year-round sex. This seems to tell us that seasons are of small importance to our reproductive success, which is probably true. If a parent must feed a child for fifteen years it really does not matter very much in what season of the year she starts the process. We can understand, then, why natural selection allows human kind year-round sex, but we must note a necessary and remarkable consequence of it: there will certainly be too many babies.

Our breeding strategy, a variant of the large-young gambit, requires that each couple arrive at the optimum size of family, not too few and not too many. But we have a sex life that must lead to a continual progression of pregnancies and births, the intervals determined only by the gestation period, and some further delay imposed on the mother as the next oestrous cycle is delayed while she suckles the baby. It is claimed that a healthy woman can give birth to twenty-five children. An inescapable conclusion follows from this: the primeval human breeding strategy involved a culling of the surplus. In order to avoid evolutionary disaster, our ancestors must have been able to call a halt to each expanding family. There is, in fact, plenty of evidence for how the halt is called in primeval human societies.

The most obvious and direct check to the baby crop is to kill the surplus, a practice we know as "infanticide." We give the practice a name because it *is* something we know about. Surplus babies were and, indeed, are, killed (or left to die) by their parents. This behavior is a frequent, persistent, and hence normal, property of humans. The modern practice of getting at the babies before they have actually been born—abortion—is but a variant on the theme. And it is clearly something that it was necessary for people to be made to do by natural selection. The same selective process that gave advantages to some distant ancestors of ours who persisted in year-round sex had to ensure that the consequences of that sex were not families too large to raise. Infanticide thus gave selective advantage to the families that were programmed to do it.

It is essential to notice here that infanticide is not a device to regulate population, neither is it a device to slow the growth of population. Infanticide, by helping to keep the family at optimum bigness, is a device to promote the reproductive success of the parents. It is the removing of surplus babies that the remainder may live to maturity and themselves reproduce. *Infanticide is a habit that tends to make the population grow.*

When people leave a baby to die, it is a more or less deliberate act. They do not want the baby or, if some human yearning makes them part with it reluctantly, then they cannot *afford* it. So they go away and leave it, deliberately. But many other customs of our ancestors, various taboos of sex and life, though less obviously deliberate, had the same effect. Sex taboo and puberty ritual worked to suppress the number of babies, tending to bring the numbers of children down to what each couple could afford. Taboos are learned behavior,

37

products of people doing what other people tell them to do. If a taboo results in regulating the number of children toward what is optimum for the time and place of the society which practices the taboo, then that society will maintain its numbers. More, the successful rearing of optimum numbers of children may tend to force the numbers up, which might result in emigration of those surplus to the adult niche-spaces available. The taboo, successful in regulating births to promote reproductive success, would spread.

The human breeding strategy, then, is based on sexual habits that lead to a surplus of babies, balanced by patterns of behavior that reduce or halt this continued accretion by culling. The methods of culling are either deliberate (infanticide) or properties of social behavior (taboos) that probably serve a number of other functions as well. But, whether by infanticide or learned taboo, these methods of stemming the flood of babies to what is convenient all result from the use of intelligence. It is the purely human quality of a developed intelligence that allows our curious sexual appetite to be a useful part of our breeding strategy.

Intelligence gives great advantage in the fashioning of an ideal family because it gives us a very good tool for working out what it is possible for a couple to do. A hen blackbird can only measure the family possibilities of her brood from how much food she can find in the early spring. The blackbird simply uses "food now" as a cue to set the size of her family. But "food now" is of very limited use in predicting the food of the next fifteen years, as a human parent must. It takes a sophisticated sensing system to measure enough parameters of the environment to make any reasonable prediction about what can be done over fifteen years. In a detailed way it cannot be done at all—remember that twentieth-century science still cannot predict next year's weather —but, by recalling the experience of the last few years and extrapolating this experience forward, it is possible to arrive at a tolerable estimate. Intelligence lets us do this.

Intelligence, therefore, must have served the first humans very well indeed in letting them refine their breeding policy in ways more subtle and precise than were possible for the other large animals around them. And it is through successful breeding of individuals that you win the great game of evolution by natural selection. Since the breeding strategy is at the front of the struggle brought about by natural selection, any improvement will give immediate returns in the numbers of your personal descendants pushing into the niche-spaces of the next generation. I suggest that refining the size of a family yields more immediate evolutionary reward than does an opposing thumb or walking erect. It may be, therefore, that one of the strong pressures leading to the development of intelligence itself was that it served the human breeding strategy so well.

This argument does not overlook the fact that people who have always lost children to accident and disease need a supply of babies to offset the losses. Such wastage always enters into the calculation of the number of children a couple can afford as, indeed, it must enter into the programs that determine the clutch size in birds or the litter size in cats. Doubtless one of the advantages of our recurrent births is that replacements can be added to a growing family to make good losses by accident. But the essential strategy remains clear. Central to human breeding is that each couple is programmed to choose the largest number of children that they can afford. It was so in the earliest stone ages, and it is still so today. Any families of people without this program have long since been taken from our midst by natural selection, except perhaps for some modern misfits on whom selection is still working.

When we lived in a constant niche like all the other animals, there were essentially no population consequences of this perfected breeding strategy of ours. As the niche never changed, so the numbers of the people never changed. There was competition for niche-spaces in the next generations, and some of the rules by which these competitions were decided are exciting to think about. Some were animal rules, like the appeal to a mate and the primitive skills of getting food. Some were human, like the resort to tribal warfare, vendetta, and blood feud—all mechanisms that served to remove surplus contestants from the Niche-space Stakes, whatever the people imagined they were gaining by these things, and whatever other social purposes they might serve. But if these habits had not chosen among contestants for a niche-space in which to live, other habits would have been found instead. The essential fact was that our remote ancestors had a superbly effective breeding strategy, but that they lived in a fixed niche. Their populations, therefore, remained essentially constant, certainly for several tens of thousands of years.

What has changed since those ancient times is not the breeding strategy, but the niche. We have learned to live not only as hunters or gatherers, but as farmers and industrialists as well. These are quite different ways of life from those of our ancestors, and they can provide for populations of quite different sizes. This is why our populations have grown since those early days: because the niche has changed. All of us still breed to press more of our descendants into the next generation than there is room for. In the old days this made no difference, because the job opportunities of niche never changed. When we started to change our niche, the opportunities for life went up, and our numbers rose accordingly.

Each human way of life will have its own characteristic size of family. This obviously applies to people living in different places, where resources may be good or bad for child rearing, but it also applies to different standards of life, to rich and poor. If a family is poor and its members live among people who are poor also, the number of children in the typical family will reflect their poverty. Bringing up children to live in poverty does not demand much preparation for the child. It will need food over the years, but only of the simplest sort, and the time for which the child will be a net drain on the family's food will be comparatively short. There need be none of those years of elaborate schooling or technical training, when the child contributes nothing to the family, yet has to be fed, clothed and housed. The poor child is cheap to rear, or, as an ecologist would say, it requires few resources. It is, therefore, quite possible for a family living in a culture of relative poverty to find the resources for many children.

The poor tend to have large families, as we are repeatedly told by those who anguish over the bounding populations of that poverty-stricken portion of the nations euphemistically called "the developing countries." These large families are fully predictable from a knowledge of the human breeding strategy. Because it takes scant resources to raise a child in poverty, the hopelessly poor will opt for large families. They are doing their Darwinian thing, estimating the number of children that can be raised to compete for niche-spaces in their world of chronic poverty and then arranging to have families of this calculated size.

The wealthy, on the other hand, must plan for each child to be able to compete for niche-space in a world of wealth. Tradition for the wealthy requires that each child be supported for more years, that it be given more things, that it have a larger home, perhaps even that it be provided the horrendous costs of Harvard or Heidelberg. When the Darwinian cost-accounting is done in a wealthy family, the stark fact is that the certain and successful rearing of a child, fully equipped to become itself a parent in its parents' world, requires a very heavy investment. Wealthy parents, like poor parents, seek to raise the largest number of children that they can afford, for this is their animal breeding strategy which has never changed. But wealthy people cannot *afford* very many children, despite their wealth.

Professional demographers, who make it their business to chart and project the course of human populations, describe the switch to smaller families in wealthier societies as "the demographic transition." Many a demographer ends a speech or article on the rising human population with this "demographic transition" as being a reason for hope. "Let us make the Third World wealthy," they say, "and the population problem will go away." But to find much hope in this is to fail to understand the ecological causes of those smaller families of the wealthy.

Smaller families for the rich than for the poor are explained and predicted by the ecological analysis of the human breeding strategy, as we have seen. But this does not mean that numbers in a rich society will not rise, only that they will rise more slowly. Breeding strategy still ensures that each couple will raise the largest number of children it can afford and, under most conditions of wealth, this is likely to be more than enough to replace the parents.

Making the poor wealthy will slow the rate at which children are raised, giving us more time to anticipate or plan the historical happenings that their crowding will bring, but it can never stop the children coming in excess supply.

It is essential to realize that people of poor countries have their large families from choice. The poor themselves will tell you that they need to have children to look after them in their old age, or to have sons to go out to work when they are ten years old, or to have daughters whose marriage will bind families together. They might even say that children are a "comfort"; that they like children. These are but ways of saying that they are looking to the number of children decreed by their way of life, or the number demanded for them by the workings of the human breeding strategy. In either language, the poor have large families because they want large families. Providing the poor with birth-control devices will not result in fewer children.

All birth-control devices must be useful to people in their attempts to fashion their optimum family. Formerly, when taboo and sexual custom had failed to arrest the baby flood at a convenient stage, people had to turn to infanticide. It is more comfortable to turn to the condom, the pill, the diaphragm and the lump of plastic perched in the uterus—all devices that let people regulate the arrival of children. These are all tools in the service of the human breeding strategy. Like infanticide itself, their long-term effect must be to increase the

43

pressure of individual families on the numbers in the next generation.

But it is still possible for the human breeding strategy to cause population losses, as well as population gains. This will happen when a community is reduced to such despair that the average opinion of the ideal size of family puts it close to zero. Or, if hope yet allows some couples to start families, then the conditions of the people are so desperate that they cannot succeed. A single generation of desperation can remove a whole community for good.

It is to this possibility of near total failure of the breeding effort, not to massacres of adults, that we must look for the decline of populations in history. Elimination of adults by disease is possible, and no one doubts the killing power of the Black Death in medieval Europe or of smallpox for the American Indians. But elimination of populations by war or massacre is much less easily believed. The German death camps and *Einsatzkommandos* of the Second World War showed how extremely difficult it is to kill off whole populations. Even with modern technology, Hitler's Germans found that processing people to death was hard work, and work for which even the most brutal were apt to lose their taste. The elimination of conquered populations with no better instruments than swords or spears is close to incredible. But people would be just as effectively eliminated from the pages of history if they failed to breed. The only deaths that waste the population then are the result of old age.

Slavery is a more plausible cause of population loss than deliberate killing, because people in the harsher kinds of slavery may fail to rear replacement families. Slaves might not care to raise children at all, or the chances of the child dying before being old enough to breed itself may be very high. So a conquered people can be removed by enslaving them tightly for twenty years.

Even more starkly, a conquered people may be removed if they are forced out of their traditional lands. A conquering army can force people to walk away as refugees far more easily than it can rid itself of the nuisance of their presence by killing them. The vanquished are "driven out," which gives an impression that people are forced to move "over there," not a nice place, perhaps, but still somewhere. Yet an ecologist must ask in what circumstances an "over there" can actually exist. Another country, yes, indeed. But if the people who are driven out are to live in that other country they must find it empty when they get there. In the real world it will not be empty, but full of people living at whatever density that other country can support. The newcomers, perhaps after fighting, might be able to eke out a subsistence "over there," but finding resources to raise the traditional number of children is quite another matter.

History books give many hints of times when peoples fade away after their lands were taken from them in conquest. In a later chapter I discuss the fate of the people of the city of Carthage after the Roman boast that they had fully destroyed the city. The world of the 1970s was given a dark example of what can happen to a conquered people when the captors of Phnom Penh in Laos drove its citizens into the countryside. It is certain that very many people, apparently several hundred thousand at least, died in the first month after they were moved, and what accounts we have of conditions where the people were "resettled" make it clear that the survivors must have little chance of raising children. The people of Phnom Penh were not killed, they were merely removed from their places of abode, an act which, if there is no relief from armies of their friends, means that they will leave few descendants in later generations.

Southeast Asia is now (1979) again showing us how a conqueror may elect to "drive out" people whose niche-spaces the victors covet as the Vietnamese send out the refugees whom we call "boat people." The Communist victors from Hanoi apparently want the job opportunities, professions, land and accumulated possessions, in fact all the best niche-space of the Chinese population, and are simply turning the people adrift in boats. Some of these people are being taken in by other countries that can scarcely afford to do so, but these havens offered to the wanderers must be a peculiar property of a modern world that has both the conscience and the technical means of finding a living for these extra people. Throughout most of history there would have been no homes for the boat people. Even as recently as a century ago, tribes of Indians were moved in like manner across parts of both North and South America, inevitably to fade away to some small portion of their original numbers, nor can the sensitive conscience of the modern West bring back their lost populations.

So the human breeding strategy, which is cost-effective and able to press descendants into later generations with an economy of effort unmatchable by mass-breeding insects or fish, is yet very vulnerable to the coming of hard times. In a few generations whole peoples can be eliminated or drastically reduced by close slavery or forced emigration. Neither process involves much direct killing. Those beaten in war can lose their posterity because they find it hard to raise children, and the only agency of death needed to take them from the pages of history is old age.

As squirrels or bears have each a packet of habits that fits them to getting food and avoiding hazard, so must the primeval human populations have had their own peculiar set of habits. They were able to gather food, to find what was good by trial and error, to learn from one another what to eat and where to find it, to band together in the hunt, to make shelters, to make sensible decisions about whether to run or to fight, to set aside food for the lean season, to go in when it rained. These traits of early humanity were all parameters of the human niche. Together they gave individual people the skills to extract an array of foodstuffs and other resources from the wild landscape in which they lived, even as they let the people escape danger while they worked their special tricks of breeding. These many traits of those times defined the ancestral human niche and the niche, in turn, set the size of the human population.

We no longer live in the ancient human niche, but we still could, or rather, some small number of us could since there would not be room for many. It must, therefore, follow that we still possess the traits that equipped us for that ancient niche, even though we have turned our skills into living in quite different ways. We have invented and learned most of our new ways, so they must be wholly new. But some of the ancient adaptations that we did not have to learn are still with us. Before pondering the changing and learned parts of our niche it is well to see how many of our doings may still be programmed remnants of a remote animal past.

It is tempting and fashionable to think of modern

people as animals still and to see in the struggles of modern societies the hand of our primitive nature. The reasoning goes something like this: "Our habits once fitted us for a wild-animal life; how much then are we animals still: acting selfish, hurting others, going to war even, as animal passions direct us?" Notice that I deliberately use the word "animal" when talking of our supposed passions, not the word "human." This is because many popular writers have deliberately compared our behavior with what they know of behavior in animals. But our ancestors were people, not animals, a very important difference to an ecologist.

Books setting out a behavioral hypothesis of history, based on studies of other animals, have a wide vogue. We are told about aggression in dogs; and then human soldiers are likened to attacking dobermans. We hear of songbirds singing over territory; and then are told of whole nations going to war because of the same imperative urge which drives the birds to territorial behavior. We even find our social lives put down as fables of an ape which has lost its hair. All these ideas, few of them seeming realistic to a professional ecologist, can be read as claims that the ancient human niche is showing through. And that niche is explicitly said to share qualities with apes, or birds, or dogs, or whatever the pet animal is.

It must be true that the actions of individual people, particularly at moments of stress, can be simple responses which once let our ancient forefathers do the right thing. But this is not to say that those simple animal responses of ours need be what we see in birds or apes. We never were birds or apes; we always were a quite different and unique species fashioned to its own special niche. We do not snarl like dogs or hold territory like birds, nor are we the substance of an ape without its fur. It is a comforting thing about niche theory that it lets us see ourselves as properly unique.

Every species has its own niche, unique and distinct, and when a new species is formed it has left behind the habits and behavior of the evolving individuals of an ancient past. The old traits that will still remain are only those which are suited to the new niche of the new species. It follows that when the simplest animal passions drive us, as they sometimes do, these drives are the drives of primitive people, not of hairless apes, still less of birds. What suits a bird to a bush or an ape to a tree is not inherently likely to have suited hunter-gatherers to the society of their times.

We can, it is true, find teasing parallels between the behavior of many animals and individual human responses to many things; to a stranger, to peril, to food. For instance, we are not the only animals who can respond to a threat of instant death by running away. Running away is a very good response to danger; it is very often an appropriate response. When it succeeds it gives fitness and is "approved" by natural selection, so that very many kinds of animal have acquired it. Other responses may be more subtle, and we may well learn to recognize them in ourselves by studying them in animals. But whether we have them or not has very little to do with the fact that apes and people had some incredibly remote common ancestor. What is important is that these animal traits of ours served people well when we lived in a niche that had many less learned dimensions than has our modern niche. We behaved in the old times as we behave now, like individuals of the species *Homo sapiens,* an animal that has little in common with an ape.

49

It is not hard to find in the aspirations of modern people echoes of our real primeval niche. We like space and horizons. We cling into social groups, of several families as well as the family itself. We gather, collect and store things. We throw stones, sometimes a piece of brick taken from a slum alley to be shied at a cat, sometimes an aerodynamic shape, made of plastic or leather, to be thrown between posts or at a bat. We fashion weapons. We live in houses, and many of us seem urged to spend time tinkering with the structure of those houses. We make clothes. Often we need to go awandering and seeking adventure, particularly our young males between the ages of puberty and the arrival of the first child. Our ancient ancestors, the people of the earliest stone age, must certainly have done all these things. They are all parts of the human niche. If some of these things are done by other animals too, that is interesting, though irrelevant to the course of human history.

But it seems certain that the most important dimensions of the ancient human niche were learned. Even the simplest of tribal lives is complicated by the standards of animal lives. Tribal peoples who have survived to our own times exist in a matrix of custom, ritual and taboo. The things all the people do, from gathering food to going to war, are done by the rules of custom; they are learned. Yet to the ecological eye all these learned things fit the people to get their livelihoods and persist as tribes down long spans of generations. The habits serve each individual of the tribe as the programmed food search of a bird species serves the individual bird. Custom and ritual define the boundaries of these tribal niches.

We know that ritual and taboo can help the family produce an optimum number of children. In a like manner can taboos shape the everyday activities of the niche. It does not matter what people, or their elders and witch doctors, thought they were doing when they followed the tenets of taboo; it is only the effect that matters. If the effect of a custom is to keep people alive, or to refine their breeding strategy, then the people who practice that custom will flourish; and people who avoid the practice may fare less well.

Ancient people learned their professions of life, just as the followers of modern professions learn theirs. It was this fact that made us ready for the dramatic changes of niche that were to come later. But that ability to learn could serve our forefathers only if it was used with caution. Learn and stick by what you have learned had to be the rule, because chopping and changing of life styles would have been fatal. The possible food supplies were very restricted and the individual had to get his share while living in a social group. Someone who kept picking up new ideas, throwing out what he had learned and learning something new, would be most unlikely to flourish. So it was essential to the ancient human niche that people not change their habits easily. Learn the niche, yes; but stick to what the elders teach you. The power to change the niche by learning, therefore, was coupled from the earliest days with a necessary conservatism that resists the change. Both power to change and resistance to that change are fundamental properties of the human niche.

Even in early prehistoric times this coupling of properties had let our species do something that no other animal had ever done before; it let us spread over most of the world. People filled all Africa, all Europe, all Asia from the tip of India to Northern Siberia, islands far to sea, and, after the Bering Strait was crossed in the last ice age, all of the Americas from Alaska to Cape Horn. All over the world our kind could carry on the trades of hunting or gathering in various combinations, because people evolved local customs suited to each vastly different place.

No other animal ever achieved anything like our primeval range, except those few, like rats and sparrows who now get their living from us. Butterflies or foxes that spread to new areas do so by forming new species and the new kind then has a niche suited to the new place. But we *learned* new niches, then fixed them by social custom. We had covered the earth well before what we now think of as history began. This spread of ours was an overture to what was to come, and it was worked by that same unique human trick of changing the niche.

This quality of learning most of the important parameters of our niche gives additional reason for rejecting views which equate the activities of human society with supposed imperative urges to territory or aggression. People have long learned by trial and error what is successful behavior; indeed, this quality of learning how to behave is the most distinguishing quality of our kind. When a tribe goes to war it does so because ancient tradition decrees war in some circumstances, not because its people were "aggressive."

Because wars had to be fought in space, it does not follow that fighting people act like mindless birds fluttering their threat displays over "territory." And because war means aggression it does not follow that the young men of the tribe are squabbling like cornered dogs for meat. As long as we have been the species that we are—at least fifty thousand years—we have *learned* our social lives, including the community fighting that is sometimes necessary.

The original niche-learning that spread our species over the whole earth in prehistoric times probably had little effect on the density of those ancient populations, because each subtle variant on the ancient human niche exploited very similar resources of food. We were always either hunters or gatherers or both. When we hunted we were rare as tigers are rare, because there is not much food to be won at the profession of big fierce hunter. When we gathered we were only less rare, perhaps achieving populations like those of bears, which exploit many of the same resources that we used. Our changes of niche let us find these foodstuffs in new habitats and we extended our range. But this was all.

The really novel event that launched the process which we call history was the habit of changing the niche to yield new forms of foodstuffs. We taught ourselves to grow food instead of hunting and cropping it. This was a very radical change of niche, indeed. It was made possible by the twin accidents of a digestive system that could handle a very wide range of foodstuffs and that vital habit of ours, already long established, of learning the dimensions of the niche in which we must live. It was as if a tiger had taken to growing corn. It was an event completely without precedent in the history of life on earth. And its consequence was a change in the density of human populations.

When an animal lives in a fixed niche its population is at once fixed also, no matter how vigorously the animals breed. This was true also of the human populations of the first forty-or-so thousand years of our existence as a species. True, we learned our niche from our elders, but it was fixed in very narrow limits by the foods available, and then fixed again by custom. Even our highly perfected breeding strategy could make no difference to our numbers then. But when we learned to change the niche in ways that massively increased the food supply we broke this pervading control of numbers by niche.

Increasing the food supply by changing the niche gave our perfected breeding strategy a chance to show of what it was inherently capable. Unless we changed the breeding strategy, which we have never done, there would, inevitably, be a great increase in the number of people living. Each individual, remember, is programmed to try to thrust as many descendants as possible into the next generation, and it is competition for niche space that winnows the surplus. But, if more niche spaces can be made almost at will, there will be no more competition. All offspring raised to maturity will find a niche in which they can live and raise offspring of their own.

The Darwinian breeding strategy of the animal that can create unlimited jobs for its offspring would lead to an unlimited number of survivors. Substitute the words "very large" for the word "unlimited" in the above sentence and we have one of the results of the great people-experiment of changing the niche without changing the breeding strategy; people have overrun the world.

When the constraint of a fixed and special food supply was removed, we were freed from many of the other restraints that fixed our niche also. With food more easily come by, it was not so necessary for each individual life to conform so closely to the tribal norm, and it was possible to train oneself to live in different and pleasing ways. The result is the generous variety of lives that we have learned to live. In ecological language, we learned to live in a wide variety of niches, each with its own set of resources.

And now we come to the great rub of history. We learn to live in fine niches, each of which requires a unique set of resources. Some of these new niches are very broad, requiring space and variety for each individual. But our breeding strategy continues to crowd in more people. As the numbers grow they can be provided with the food parameters of the new niches easily enough, for the productivity of the whole earth has been turned over to this one species. But there are always critical densities of people for which the other parameters of any given niche can no longer be supplied. Then the next generation must live in still newer and less broad niches. This means that, far from niche setting the numbers in our population, it is the size of the population that now sets the size of the niche.

Our numbers have escaped the restraint of niche; niche is now under the restraint of numbers. This is the new fact that underlies the events of our history. In centuries when technical advance is slow, adding more people always works to deny opportunity to some as the finite cake of resources is cut into smaller slices so that each newcomer is given niche-space in which to live. This process can go on until each niche-space is so narrow as barely to support life, a condition we recognize when we talk of extreme poverty. There have been societies in history who have crowded so many people into the lands they own that most are poor and very few enjoy the ample lives of a broad niche. Something like this probably happened in the ancient Asiatic kingdoms when they became stagnant, with steep social pyramids and grotesque differences between rich and poor.

When rising numbers continually make niches smaller, we should expect the compression of existence to that of agricultural poverty to be very rapid, but in fact the process is usually gradual and prolonged. This comes from another consequence of our being able to change the niche at will. We invent new resources to support the expanded niche, just as we invent new ways of feeding. We call this process of creating more niche-space "technology" and "industry." The result is to provide large niche-spaces to larger numbers. But since the earth is finite, there is a limit to the niche-space that can be provided in these ways too. We can defer the final poverty but not avoid it.

It is clear that the human niche must be composed of two sets of dimensions, one set fixed, the other set learned. The fixed dimensions are those primeval human appetites that we have: the need for shelter, the fear of danger, the love of adventure. Since the programs for these things are written in our genes, we cannot achieve happiness without providing for their expression or release. However, the dimensions of our niche, which we learn can be compressed or stretched without affecting our future as a biological species, concern very closely what we call "the quality of life."

We alone of all animals can choose the niche in which we want to live. How shall we describe this desirable human niche? Our definition should include the dimensions that we should all like to see in the niche-spaces we occupy, but it should also be sufficiently general to allow for much variety in our individual tastes. I suggest that a gathering of American revolutionaries wrote down such a definition two hundred years ago. Their words can be rewritten to say that the human niche is bounded by a set of inalienable rights, among which are life, liberty and the pursuit of happiness.

When we describe a broad niche in terms of life, liberty and happiness we must also note that this niche, wherever it has been found, has always been lived in by people of unchanged breeding strategy. This means that numbers will rise to compress the dimensions of the niche. The food dimension is certainly the last of the three rights to be squeezed, and life can be more easily granted to dense populations than can liberty or the pursuit of happiness. A desirable human niche is, therefore, at the mercy of an unchanged breeding strategy. It is this that has caused the major events of history.

CHAPTER FOUR

WHY HISTORY HAPPENS

THE PROPER study of history begins in the Pleistocene epoch during the time that continental ice sheets reached their fullest extent. People had already lived in the world for many tens of thousands of years, fully modern people, biologically our kind. But the inventions, the social experiments, the great conquests and the flowering of intellect, which is the stuff of historians, all lay in the future. The development of our affairs before the last coming of the ice mostly concerns scientists: the students of Darwin and anthropologists. But when the ice went away again, the vital changes in human culture started. The last ice age is the great divide, and they who would truly understand the human story must begin by learning something of the people of those days.

We can piece together much about how the people of the ice age and before lived, from the marks they made on the land even then; from the shapes of their houses, from the stone weapons they left behind and from their mounds of refuse. Study of these traces tells us of people who lived by their wits, sometimes hunting big game, sometimes gathering fruit, sometimes fishing, sometimes building settlements, sometimes wandering with the seasons, sometimes changing to the food of the moment, sometimes refining techniques for settled ways of life and ordered society. But, to an ecologist, all these variations on the theme of ice age man suggest much the same thing; the resources available to these ways of life were never very large. When people hunted big game they fed as active carnivores at the very tops of their food chains. They were accordingly, and necessarily, rare, as tigers are rare. When they gathered vegetables and fruits in productive places of genial climate they should have got somewhat more to eat, but even then the pickings may have not been very great. Fruits and vegetables are seasonal things and people have to live, and to raise their children, through all seasons. These people who gathered food, with a little hunting on the side, were like bears, animals a little more common than tigers but not much so.

Made rare by their diet, the ice-age peoples were otherwise very like us, living in houses, sewing their clothes, setting aside provisions in good times for the bad times which were sure to come, loving their children, husbands and wives. Some may even already have had that commensalism with dogs which we have not been able to shake, as the sidewalks and pavements of modern cities so eloquently show. The ice-age people no doubt fought occasionally, when individuals lost their tempers, as people will, and when tribes found it suited to their policy to war on other tribes. We must expect that for some peoples the niche included fighting, that a certain amount of war, vendetta or blood feud sometimes provided necessary social ties, or helped to keep family sizes down to what the environment would support. But such fighting would have been an intermittent and essentially stable business, with the bickerings going on for generation after generation with very little being changed. There would have been no wars for conquest or empire such as is the stuff of history books.

In the family women and men of the ice age, wearing clothes, in a house, among their children and their dogs, we can see all that is essential of ourselves. The big changes are that we have become immensely common when once we were rare, that there are now rich and poor living in different ways, whereas once all people lived the same, and that we have done things such as launching wars of conquests and waging revolutions, unthinkable to ice-age peoples. Neatly tied with these changes must be the results of our technical cleverness, of the things we make, of the cities in which we live and of the ways in which we have made the land produce to feed so many people.

Both the changes in numbers and the deadly changes in habits directly result from our learning to change our ways of life, our ecological niche, almost at will, without ever changing the ancestral breeding strategy of letting every couple raise the number of children it thinks it can afford.

About nine thousand years ago some of us began to herd animals instead of hunting them, and others began to grow the first crops. These changes of niche fashioned the human destiny. They let our numbers grow.

Herding animals gives you a much better yield of

meat than hunting does, because you stop the other predators from getting any. In the old days of hunting we had to compete with cats and wolves for our prey, but herding kept all the prey for ourselves. Furthermore, the herdsman kills his beast when he wants it, not just when fortune favors the hunt, providing for a prudent use of the available food. Changing the niche from hunting to herding, therefore, meant that there would be room for many more people to live.

But the habit of growing crops was the larger and more decisive change. The first farmers gave up the role of carnivores in the ecosystems of their times and took to herbivory as if they were cows or rodents. It was this habit of agriculture which let our numbers grow from the rarity of bears to something like the commonness of rabbits.

Big, fierce animals that eat meat are always rare. They get their food, like all living things, from the energy of the sun, but they get their share only after their calories have been passed from mouth to mouth down the links of a food chain. Plants take energy from sunlight with an efficiency of only one or two percent. All the plant eaters together, those whom we call herbivores, only take their calories from plants, with an efficiency of about ten percent, so the food calories running about the earth as meat are just some portion of the ten percent of the two percent of the solar energy taken by plants. Big predators like wolves or people can get only some small fraction of these calories running about as meat for their own use, and in fact, they seem to be inefficient at gathering even these few available calories. I have recently calculated the efficiency with which a pack of wolves living on the moose herd on Isle Royale in Lake Superior gather in their meat, and I find that the terrible wolves do very badly, taking home only about 1.3 percent of the meat calories theoretically available to them. It follows that a big herbivorous animal can be about a hundred times as numerous as the big carnivorous animal that hunts them. When we gave up hunting for farming, we turned from being carnivores to being herbivores, at once offering ourselves the chance to be vastly more numerous. We broke the primeval restraints on number and marched down a food chain to where the energy was.

Not only did herding and agriculture vastly increase the quantity of food calories available to people, but they also made the supply more certain. There was less boom-and-bust to living, and death by starvation in the lean seasons became less common. Moreover, the knowledge of how to farm would let us respond to shortage of food by farming in still more novel ways or by growing still more novel crops. We could be numerous like cows or rabbits and then, when the cow and rabbit food gave out, displace caterpillars, sheep and birds, taking their food too. The ingenuity of nine thousand years ago showed us the way to cope with any shortage of food by some new trick of agriculture, until one day it will be possible for most of the plant production of the whole earth to be used to provide calories for humans.

The human population has risen these nine thousand years as the breeding strategy responded to the optimistic possibilities for family life offered by recurrent gluts of food. Our breeding strategy was, in fact, nicely preadapted to the arrival of times of good feeding. Large families became possible nine thousand years ago and all that was needed to bring them about was for people to abandon habits of restraint. They did not need to think it through; it was enough for the young to rebel. Harsh practices, which once had set the family size to what the average couple could expect to support, would now have nothing more behind them than tradition. Generations must grow up impatient of ancestral feuds, disgusted by such practices as exposing babies, daring to experiment with sexual taboos. Changing fashions in behavior manipulated the family size to circumstance, and each family got bigger.

We have no record of what went on in the societies of nine thousand years ago, but we have good knowledge of the consequences. Before the invention of agriculture people were rare; afterwards, people began the geometric increase which has resulted in our present abundance. Since this is what happens to any Darwinian species when there is a large increase in supplies of food, the only real novelty in the rise of the human population is that the necessary glut of food was provided by a change of niche. The unchanged breeding strategy did the rest.

It is revealing of the peculiarities of people that the new breeding effort came from social change. People learn from other people what family and sex habits are good for them. And when what the old folks say no longer seems right, tradition changes; though not at once. First the old tradition weakens, then the young rebel, and then a fresh generation establishes a new tradition. Unlike other animals, we can change our social habits to fit ourselves for new niches and other times, but the process must involve social unrest, the pitting of generation against generation; even sometimes armed rebellion.

We have been seeing this process in action recently in the affluent cities of America and Europe. Young people have been raised in comfort and security on a scale unknown to their parents and grope toward new life styles; and the result is a sudden gulf between young and old which we call the generation gap. The generation that grew to awareness in the 1960s, filling our universities with unrest, lived in niches quite different from the niches of their parents' youth, and they had to rebel at old teachings as they worked out the new living. Social unrest must follow if society and training do not match. And the generation of the sixties found a new sex style also, even as they groped for a new niche, just as their remote ancestors must have done when they invented agriculture.

Nine thousand years ago, no doubt, the changes were slower, but the effect was the same; to develop new life styles suited to the new resources, as well as new sexual and social customs which fashioned the size of families to the new circumstances. Since life was still simple and food no longer limiting, families would be large. This is a prediction of ecological theory. The history of our population from those early times gives confidence that the prediction is correct.

It was only when the last ice age was over that agriculture and herding led to the rise of the human population. The last gasp of late-glacial events had ended about ten thousand years ago across most of the earth, leaving the earth's climate roughly as we know it. There were still some minor movements of land, sea and ice, but for the most part the earth had settled into its present condition. In the few thousand years that followed, completely unrelated cultures in many parts of the world invented agriculture, arguing strongly that there is a link between the end of the ice age and the grand experiment of changing the human niche.

People are ice-age animals. They passed the first, and by far the larger, part of their history in the ice age. It was for ice-age life that we originally evolved as the species we remain. This is a most important fact; moreover, it is foreign to many a modern prejudice. We tend to think of ice ages as something frightening, times of cold and misery. This is completely wrong. Evolution suited people to ice ages; it follows that ice ages are suited to people.

What we now know of the climate of the last ice age makes this shocking conclusion of ice ages being good times much easier to accept. The patterns of land and sea, summer and winter, of the ice-age earth would not be at all uncongenial to our modern tastes. Most of the world had no ice on it, for the glaciers sat on northern land which is cold even today, or on high mountains where we can still find smaller, remnant glaciers. The sun still shone; the tropics were still tropical; the equator was a torrid zone then as now. The really big changes over much of the earth were the spreading of the great plains, grassland and savannas. The ice age was a time of lower rainfall on many a continental basin, with savannas where the tropical rain forest now sits on the Amazon and elsewhere, and with prairies stretching far and wide across the lands closer to the northern icecap. Furthermore, there was actually more land in the temperate and warmer latitudes than there is today, because the glaciers forced the seas to retreat. A great icecap is to be thought of as a giant ice cube taken out of the oceans by rain and snow, and kept, in cold storage, on the land. The oceans shrank to make this ice cube, so much so that every sea less than three hundred feet deep became part of the continental plains. An ice age actually makes more land in the latitudes where people like to live, good land, with warm, dry continental summers, supporting grass rather than forest, a proper place for hunter-gatherer peoples to be.

So the ice ages were the good times; the best of times, when we learned our trade as a species with natural selection as our teacher. But the good times of the last ice age came to an end at the last and there were a few thousand years of affliction through changing weather and disintegrating landscapes. Then the world settled into the pattern we now know, when weather, countryside and food supplies were all different from anything experienced before. In the starkest of ecological terms we can say that the number of human niche-spaces had changed, which meant direct and immediate adjustments to the size of the human population. In every culture everywhere, it would also follow that the numbers of children that a couple could afford to rear altered, so that a relentless pressure for new habits of fertility pressed upon the people. In the centuries of change after the comfortable normality of the ice age, human numbers everywhere must have gone up and down in synchrony with environmental change. Social mores were under constant pressure to adapt peoples' breeding strategies to an ever-changing resource. It may have been the most turbulent period in the history of our kind, unrecorded though it is.

Food limits tended to be pressed downward, since the years in which the ice sheets melted were years in which ancestral lands and sources of food disappeared before the new, postglacial order. Traditional food may have increased locally, but certainly not globally. For animals of fairly fixed habit, which people then were, change in physical surroundings is not likely to mean that you are better off. The years of glacial retreat must have been

years of human privation, however much we may talk
of "warming" as if of a good thing. If the numbers of
people changed in those years, it was probably in a
downward direction.

But when the glaciers had gone, the climate of the
world settled down into a new pattern. Things were
stable once more, though not so suitable to the ancient
human ways as the old ice-age stability had been. And
the people who faced this new order had gone through
a long trial by change. They had learned more than
ever to change their niches by experiment and learning.
They must surely have learned to change social habits
in ways that would serve the old fixed breeding strategy.
They had, indeed, learned that change itself was a suitable response to stress. The lesson was worldwide.

Given that much of the human niche is learned and
that many fertility habits too are learned, the response
of inventing a new way of life after going through a
period of stress is inherently likely. All that was needed
for hunters to turn to farming was the necessity for still
more change, together with suitable plants or animals
that could be made to do what people wanted. The ecological hypothesis perhaps cannot predict that an intelligent animal which learns its niche will invent
agriculture, but it can certainly show in what circumstances the invention is likely. Agriculture was likely
after the last ice age. We find that it happened all over
the world at that same likely time.

But humans changed the niche to farming and herding without changing their breeding strategy. The family size in every culture was accordingly adjusted to the
new certainties of food. And the human population
began its geometric increase in every country indepen-
dently. It was a great Darwinian success story, because

success, in an evolutionary sense, is measured in numbers. When we invented agriculture we began to start winning in the numbers game.

We must think that our most perfect evolutionary triumph would be a society of agricultural peasants, sedentary, marvelously numerous, living in a landscape set by the requirements of the food plant, making do with the very minimum of animal food, freed from the threats of predatory or competing animals, and having a family size again brought down to meet the needs of replacement and set by the fact that there should be no food to rear more than two or three children per couple. Peasants such as these would be the ecological apotheosis of humanity.

But people have not been able to change the human niche so completely as required by this triumphant evolutionary nightmare. They have not wanted to be the perfect food-raising food-consuming peasant. Many individuals resist peasanthood very strongly indeed, trying to preserve more ancient ways of life and even wanting to do things that the ice-age peoples could not do; they want to go adventuring like a hunter, to paint, to craft, to make machines. Furthermore, the growth of population tended to be in communal patches, to be built around social organizations which were relics of the primeval human niche. We have always been social animals. Inevitably, these social traits direct agricultural niche-living toward dense human settlements. The ideal of everyone a humble peasant was frustrated from the start by the crowding into settlements; settlements require government.

Our primeval niche let us take kindly to government, because the old social life involved divisions of labor. Hunting in groups needs collaboration and mutual support. Even herding, which ties people to beasts, requires some directed collaboration, and agriculture ties people to ground and food plant so that government for any society more dense than a one-family plot is essential. The institution of government did away with the nightmare of people being reduced to perfectly equal peasanthood. But escape may be only for the fortunate few —the governors.

The need for government in dense communities did more than just save a few individuals from the worst consequences of our change of niche. It also allowed further increases in the carrying capacity. Government could ration, distribute and hoard. Surplus and deficit could be balanced from place to place, and from season to season, ensuring an even flow of the necessities for life, making the luxury of large families the more safely enjoyed.

Our oldest book, the Hebrew Bible, gives us an idea of the importance of these early organized societies. We read of some of the children of Israel undertaking a journey to Egypt in time of famine. The Israelites of which the record speaks were still, in part, wandering herdsmen likely to suffer want if the grass failed. And some of them, worried for the safety of their children in a time of famine, went to a powerful state, their neighbor, because "there was corn in Egypt." The Egypt to which they journeyed was a city-state with magazines of grain.

The organization that we call a "city-state" is the logical, indeed the inevitable, outcome of the invention of agriculture by an animal of social habits. Agriculture requires settlement. An unchanged breeding strategy makes that settlement dense. Government in a dense community requires specialization. And a dense settlement containing both rulers and ruled must inevitably divide up the country into land to live on and land to farm. The city-state has emerged, along with a rationale that requires people within it to have different specialties—that is, different niches.

The organizers in a city-state, be they governors, bureaucrats, businessmen or priests, led active, wide-ranging lives that needed many resources; an ecologist would say that they had a broad niche. The mass of the people needed much less, little more, in fact, than would be wanted by that ideal agricultural peasantry; they had a narrow niche. This organization of life did extract more resources from the living space than a system of every family for itself, but it necessarily meant that a few people had shares larger than those of the rest. The broad niches of the governors meant wealth, but then the narrow niches of the mass could be given a new name, "poverty."

"Wealth" and "poverty" are but names we give to two extreme kinds of ecological niche. The niche of wealth demands more resources per individual than does the niche of poverty. Wealth even takes more food, for a wealthy person actually eats more calories than does a poor person. Even more importantly, the wealthy person tends to eat higher on a food chain, requiring more meat. This means that any patch of real estate probably can feed between ten and a hundred times as many of the very poor as of the very rich. How many rich people there can be, therefore, depends on how many people are trying to get their living from the land; it depends on population density. This is true even when only the single niche dimension of food is measured.

When you measure the dimensions of the human niche other than food, the contrast between possible numbers of wealthy and poor people becomes even more marked. Wealth is always associated with organizing, with ruling, with power. Those living in the wealth niche, therefore, can ensure that other desirable dimensions of niche are available to them, often including many of the more ample ice-age habits. The wealthy acquire open landscapes to live in, and their young tend to go adventuring. Even in modern times those who can afford it tend to hunt or ride horses, and these things have been main preoccupations in the past. The wealthy of every society of which we have record have built the free use of land into their personal niches, strong evidence that we cannot throw off completely the aptitudes we had for hunting and gathering in our original home, the balmy earth of the last ice age. But these are aptitudes possible only for people living at low density. They require very many resources of the real estate per individual.

The immense flux of resources required for each niche-space of wealth can best be realized by reflecting on just one propensity of the wealthy, the propensity to choose. The wealthy seek variety, both in daily activity and in real opportunity. But any freedom of choice must mean that, for everything done, there be something left undone. Freedom and wealth, which are to some extent linked, require very many resources per niche-space. The wealthy, and the truly free, therefore, must be rare.

Wealth and poverty are both inventions of agriculture-based humanity, but poverty is more of an invention than wealth. We make people poor by denying them the types of food, activities and space that were consumed in the primeval human niche, whereas the wealthy retain many of these old assets.

A young city-state could provide very well for its people. The wealthier classes not only could enjoy satisfying lives themselves, but also could hope to improve the lot of the mass through the invention of agriculture, irrigation, magazines of grain, and the rest. By multiplying the original carrying capacity many times with their city-state technology, larger niches (or wealth) could, in theory, be provided for all. This being so, and taking an optimistic view of human nature, we can sketch out the likely sequence of events in the developing state.

Good people worked to help the poor, and with high hope. The earth was a bountiful place whose wealth people were learning more and more to exploit. All that must be needed to abolish poverty seemed to be diligence and good will. Well-intentioned governors were diligent, and their efforts yielded more and more resources which could be spread among the people. And the efforts of good will were massively supported by the efforts of greed, as the people of commerce used their cleverness to extract wealth from the land. These efforts always should have yielded enough resource to abolish poverty, but they never did; the unchanged human breeding strategy saw to that.

Every couple, rich and poor alike, continued to rear as many children as it could afford. Numbers always rose. The extra resources wrung from the land by cleverness and industry always went to supporting more people at the old levels. As fast as a few individuals could be raised out of poverty, as fast as the actual numbers of people living richer lives increased, so also more babies were born into the world to swell the actual numbers of the poor.

Population growth is a geometric, or exponential, process. The cleverest of people, and the most enlightened of governments, have never increased the flow of resources exponentially at an even faster rate than the growth in demand represented by the extra mouths, except for short periods of rapid technical advance. Industrial societies of the West are experiencing one of those short periods of rapid advance at the moment, and there have been others in the past. But always a plateau has been reached. It must be so. The rate of increasing production falls but the rate of population growth does not fall. Then poverty must get worse and more visible, for not only do the numbers of people who must be poor increase, but each poor family finds itself poorer and poorer.

Every ingenious and enlightened society has, at the last, found itself facing this grim permanence of poverty. To their best people poverty has seemed a demonic scourge that would defy their efforts forever. We hear even the voice of Jesus of Nazareth calling out from the crowded triumphs of part of the Roman Empire that the poor would be with us always. But the poor have always been with us up to now only because people have never willingly modified the animal breeding strategy of our ancestors. Any enlightened civilization which had accepted breeding restraints could have banished poverty, but none has done so because none has realized that its large and happy families were the real cause of its poverty. Ecology's first social law may be written, "*All poverty is caused by the continued growth of population.*"

And yet a noisy propaganda is about which denies that rising populations cause poverty. We are told by most eminent politicians and international experts that the rising numbers, far from being a cause of poverty, are in fact a result of poverty. Those who think it sinful to urge small families onto others are particularly apt to

73

repeat this propaganda. "Poverty is the cause of large families" they say. "Do away with poverty—by foreign aid or by giving to charity—and the population problem will take care of itself." It is an appealing, comforting hope; but it is false.

People who lean on this propaganda are deluded by the very true observation that the moderately affluent have smaller families than the comparatively poor. They say that giving poor villagers of the Third World the money and education of someone living in a French or American suburb would result in their having smaller families, as indeed it would; provided that the new affluence was safe for a generation or more. The poor villagers would pass through a "demographic transition," as I explained before. Affluent couples cannot *afford* as many children as can poor couples because many resources are required to raise a child to affluence. The best example of this is that the rise to affluence in the modern West has been accompanied by a fall in the average size of the Western family. But this does not mean that poverty causes population growth; it is the growth that causes poverty, and the affluent West can lose its affluence by packing more people in. Poverty is growing in inner cities already.

It can happen, and often does, that populations grow more quickly in poor countries. But this does not mean that populations do not grow in wealthier states as well. In fact, we know that they do. The actual rate at which the population grows is a function of family size and hence of wealth, but it is not the rapidity of the growth that forces people to live in poverty; it is the fact of the growth itself. What matters is the eventual population density. It is the number of people per unit of resource that determines the size of a niche and, hence, what we call a standard of life. Coping with more people in each succeeding generation is the ultimate drive for technical innovation. New ideas can come either from the good will of those with broad niches already or out of the self-interest that seeks to preserve old familiar life styles. But poverty will always be present, because any large increase in resources produced by new technology will be taken up within a few generations by the provision of more poor people.

For the early stages of the growth of a civilization, therefore, niche theory predicts life in settlements, continually rising numbers, a ruling class living in broad niches that include many dimensions of the primeval human niche, technical innovation from those who have broad niches already, the persistence of poverty, and an actual increase in the numbers of the poor.

In city-states, and in later societies alike, the poor became part of the accepted scheme of things; a constant curse which would grow as numbers pressed upon contemporary resources and techniques. In the early days of any state or social system its leaders, ebullient and self-confident, would think that they could lessen the curse with time. I choose to believe that there are always, among the people who are better off, many who, from duty or compassion, work to improve the lot of the poor, unavailing though their efforts have always been. But as the numbers continue to grow, the richer classes themselves will feel the pressures on resources; their problem will be not "how can I help the poor," but "how can I see to it that my own children can live as I have lived."

There will be pressures on space as the increasing numbers of people require more land for living and growing food. There will be pressures on government as increasing numbers brought up to middle-class ways try to find space and resources for middle-class life. It will seem that theirs are troubled times when the traditional and comfortable way of doing things is threatened.

These troubled times for the ruling classes are inevitable. A broad niche requires numerous resources; an expansive way of life can be provided for only relatively few. But more young people equipped to live in an upper-class way will keep coming in succeeding generations as our breeding strategy manufactures more people trained to the new niches. There must, therefore, be competition for niche-space in that part of society living at the better standards of life. Niche-theory predicts, therefore, that rising numbers will always cause trouble for the wealthy before they cause trouble for the poor. Politicians nowadays talk of "the population problem" as if it were mainly a worry for poor nations and the underprivileged, but this is wrong. The wealthy are the ones to be squeezed because the wealthy use the resources that the new crowds will want.

When those in power must lose privilege because the numbers of their own kind rise, social unrest must follow. Social unrest, therefore, is a necessary consequence of changing the niche without changing the breeding strategy. Furthermore, the unrest will be a middle-class phenomenon and probably episodic. The troubles come from trying to pack in a few more of the relatively wealthy, not from packing in many more of the relatively poor. There is almost always room for another poor devil; but not for another successful merchant, professor, priest or senior official.

I suggest it is axiomatic of human history that social

upheavals, even revolutions, do not emerge from the ranks of the poor, for all the claims of Marxists that they do. They come from disaffected individuals of the middle classes, the people who experience real ecological crowding and who must compete for the right to live better than the mass. In the armed revolutions of Anglo-Saxon states there can be no doubt of this. It was the barons of King John who forced him to sign the Great Charter, the London merchants of the Parliament who revolted against Charles, the mercantile classes who put William of Orange on the throne, the eighteenth-century gentlemen who founded the United States in battle. In France, it was the refusal of the governing aristocracy to make room for the new wealth forged from trade and manufacture that broke the possibilities of orderly government and led to revolution. Even in the Russia of 1917, it was the long agitation of "intellectuals" to remake the country in their own image that made revolution possible when Russian armies had suffered ugly loss.

The episodic quality of these revolutions comes about because scattered disaffection alone may have little result. Individuals can wage a brief struggle for the niche of their parents, then accept defeat and sink to a narrower vocation in life.

This is surely what happens to most people surplus to the more desirable niches most of the time. But unrest or revolution requires a critical mass of the disaffected. The first German war provided such a mass for Russia, and a lost war can often multiply the numbers of the frustrated in this way. Yet the more fundamental cause of all revolutions is increasing technical competence which lets more people than before aspire to the broad niches of middle-class ways. Economic growth yields more room at the top; population growth causes even more people to want a higher-class life; the two processes combined produce the critical mass; and revolution follows.

Middle- and upper-class niches are inventions. They are developments of our original trick of changing the primeval niche through agriculture and settlement. When new niches are first invented, few people live in them; an ecologist would say the niches are "empty." We should expect, therefore, that many generations must pass before life in these new niches could be crowded. It must follow, then, that the social unrest, which is the prime indicator of crowding in these better niches, will always be long in coming. Lulls of social peace occur, particularly as a small inventive state begins the process of growth. Social unrest follows, always as a distinct episode. This is why revolutions are revolutionary, a sudden upset of the old ways, as in France and Russia, or the upheaval of 1848 when kingdoms collapsed all over Europe. These events all followed technical change and rapid population growth, but were decades in the making.

It is, of course, true that recruits to middle-class niches come from trained youths from niches below, as well as by the reproduction of people who are already middle class. This can make for extremely rapid filling of any new niche-space that can be invented. But training people to niches of affluence is a property of very advanced, indeed of modern, societies only. Many modern problems come from this, like the troubles we begin to feel as we thrust up to half our populations through universities only to find that there are not enough available jobs that need or can use university training. Our economies, not the output of universities, set the numbers of broad niches available, and the unrest of the 1960s is probably but a first taste of the consequences.

When a country starts on mass education even before there is a rapid expansion of the niche-space through technology, as many in the Third World are doing now, the result must be a social crisis. The crisis is like the excess production of aeronautical engineers, which I described earlier, but on a national scale and for all the appetites of middle- or upper-class life. In a version of the old saying about more chiefs than Indians, it is a deliberate production of more chiefs than there are chief jobs available. The only escape for the surplus of the newly educated in one of these countries is emigration, if some more developed country will take you; the only escape for the government is repression of the new intelligentsia. The developing world is rich in examples of both these measures.

But deliberate recruiting to the middle class will have had only modest importance in other civilizations and in earlier times. Crowding people into the more desirable niches has caused most of the more striking episodes of history—the wars of conquest and trade, which I describe later—but this crowding largely came about by recruitment from the children of those already in the better niches, not by training people from the poorer niches below.

Crowding in the upper ranks must produce a response in the government of society. We can expect that the descendants of those who once labored for the poor might well become inward looking, concerned only with the defense of their own privilege. Like poverty itself, a gradually repressive ruling class must be the inevitable consequence of indefinite population growth.

Ruling classes that feel themselves threatened by the

social pressures of a rising population have only two courses of action open to them. They can find more resources to provide good niches for more people or they can restrain the pressure on niche-space by a system of oppression. The most interesting ways of increasing the flow of resources include trade, colonies and war. These are always tried. The alternative, constraining the appetites of rising numbers by some system of force, is also always tried. It involves regimentation, bureaucracy, class, rationing and caste.

Oppression carrying the force of law is an ancient human experience. But for an organized society, with division of labor and people living in many niches, the actual institutions of constraint cannot be so simple as the phrases "social oppression" or "tyranny" might suggest. Society must allocate people to different walks of life, not just hold them down. What evolves then is a caste system, the best-known being that found in India at the time of the British conquest.

In the old India people of high caste had ample lives, civilized and cultured, requiring many resources. People of the lowest caste were at bare subsistence, denied even the right to live in villages. And there were ranked castes in between. These separated castes were closely comparable to the separated niches of different species of animal. Each had its own way of life; each required special circumstances for its inhabitants to flourish; each must have developed a characteristic pattern of rearing the young. People of each caste probably also had a characteristic clutch size (I have no data, but it is likely to have been so), and people of each caste certainly had developed mechanisms which restricted breeding with people of another caste just as breeding is restricted between biological species. The caste system fixed niches, and it fixed most of them small.

An essential quality of the Indian castes was that they did not just rank people in lines of status; castes also defined jobs in society. People were restricted to caste, but they were also given the opportunities of caste. The names that Western scribes gave to castes were not always accurate: "washermen" did not just wash, and "barbers" did not just cut hair. The caste names can often be traced to a function given people of the caste in ceremonial religion, but the crucial fact was that each caste had a distinct role in society. You knew your place if you belonged to a caste. Perhaps more importantly, if you belonged to a caste, you knew that there *was* a place for you.

Castes promote a stable society because they ration people to jobs, not all of which are the most desirable. They ease the pressures of crowding on the broad niches of the most cultured. Furthermore, castes are logical for an animal who maintained its primeval niche by learned taboo. The members of ice-age tribes learned their niches and fixed them in their descendants by the mysteries of tradition. Likewise, the members of each caste pass on to their children the mysteries and traditions of the caste which will fit each child to raise children for the caste niche of a later generation. Caste is both profession and niche, and the social system of caste rations people to niches. The system oppresses people of low caste as effectively as repressive law, yet it needs no policemen.

A civilization with castes survives because it avoids some of the social unrest that would afflict a society proclaiming that all its people were equal. But, even so, it must collapse if the pressure of rising numbers within each caste continues to grow. Then each caste, particularly a high-status caste with an expanded way of life, would suffer its own crisis of being unable to make room for the next generation. One way out for the high castes is to prevent some of the newcomers from breeding through a celibate priesthood, nunneries, or other social bars to parenthood like requiring large dowries before the young can marry. Another way is to let surplus competitors for caste space lose caste, to bump them into the caste below, whose narrower niches can stand the crowding better. This apparently happened in India, where losing caste is well documented.

Things may be simpler in the very lowest castes, because it may always be hard to raise large families there. It is quite possible to have people, outcasts of society, living so miserably that a couple can rarely raise even a replacement family. Births may always trail deaths in the outcasts, if their poverty is extreme, and the population losses in the outcasts could very well reflect the crowding caused by accepting rejects from more fortunate castes above. With a large enough caste for the desperately poor, therefore, a system might very well regulate the population of a whole nation by balancing the gains in high caste children with losses in the low; producing a stable society, though scarcely an admirable one.

The records of British India give broad hints that stability through caste was in fact achieved in the above way. Certainly there were many people in the lowest castes, and they often had a hard time raising children. Infanticide was so common a practice that the first British conquerors were horrified and devoted much legal indignation to stamping out the practice. Evidently families had to be so tightly regulated that even this barbarous means had to be used in civilized India. There was a particular emphasis on killing female babies, something that makes good sense in hard times because boys could help to keep a family without the risk of increasing it.

It is probably by ejecting people from castes, or by forcing some to be childless, that the ancient Oriental civilizations persisted so long, crowded in each walk of life, the numbers in crude check by side effects of social behavior that the people did not understand, oppressed in the sense that everyone was rationed to a way of life they could scarcely help, all raising the optimum family as worked out by their caste taboo. It was the stability of neither change nor opportunity.

The English conquerors were appalled by the Indian castes. They thought them an ugly oppression, a living rebuttal of the English rhetoric about liberty. And yet they had their own caste system, less rigid and more benign though it was. The lowest English caste of laborers lived ample lives by comparison with Indian peasants—at least before nineteenth-century industrialism —and they practiced nothing as nasty as infanticide to keep families small. Instead, they shipped the surplus people to new lands overseas, lands of low populations of hunting or gathering peoples, lands seized by British arms. The British had successfully used the solutions of emigration and conquest to mitigate the severity of social caste. Were it not for this, a Georgian laborer would not have lived so well that people are now prepared to pay $50,000 for his cottage.

The disgust of Europeans at the "Asiatic" practice of infanticide, however, may have been a little ingenuous. In Christian Europe before welfare states, people left surplus children to be collected by the Church. The Church did its best, but its best was not very good. Most of the children died before they became adult, so that "angel makers" was a good name for Church orphan-

83

ages. "Infanticide" is an ugly word; a "foundling" is better. Both were ways of eliminating children of couples desperate to keep the family size down to what they could afford.

But remember, neither infanticide nor leaving foundlings is a device to restrict the numbers of adults, or to regulate the population. Each serves the breeding strategy of raising the most children possible. Only when what is possible is less than the replacement number will the growth of population be checked in this way, and then stopping population growth is a side effect not intended by the parents. At the bottom of a caste system of an old civilization social misery may be so great that this low number of possible children is actually reached.

Castes have been described from many ancient societies—Egypt, Greece, Rome, Persia, Fiji as well as India and Europe of past centuries. Castes are apparently ubiquitous. They ration resources among the populace when broad niches are not attainable by all. They raise and educate an individual to one only of the many niches of a society.

Caste systems must largely fail before universal education, which, at least in part, trains people to choose from a variety of vocations for which they might be prepared. But education does not remove the need for constraint; if caste no longer works to choose a niche, some other constraint has to be invented. In a market economy, the individual is allocated a niche by economic circumstance. In a socialist state the individual is allocated a niche by a government official. But people are still sent to a way of life that has to be, for most of them, less than the best. It is always crowded round the broader niches, and the more dense the population, the more crowded it will be. The defenders of a high way of life must push against the competition. *Social oppression is an inevitable consequence of the continued rise of population.*

After the original inventions of wealth and poverty, therefore, niche theory predicts:

That middle and upper classes will be the first to feel the pressures of crowding.

That ruling classes which previously were sympathetic to the mass will become selfish and oppressive.

That social troubles will be episodic rather than continuous.

That methods of allocating people to the more narrow niches will evolve. Caste systems are the most human of these methods, but capitalist economies and socialism have their equivalents.

That even under oppression, populations will be stable only if the optimum family for the most miserable class is less than the needs of replacement.

When you run out of niche-space for the good life, you can always look for more somewhere else—through trade, through colonies and through aggressive war. We think of trade as "good," colonies and conquest as "not so good." Yet all three serve to tap the resources of other people's land. And they all need military hardware for success.

Trade is the simplest of the three ways to expand. You stay where you are and fetch objects you want in ships. This gives you the very great benefit of doing your adventuring away from home, and it gives a broad niche to the traders; this is the first and most immediate consequence of trade. Life for the traders is, indeed, the motive for trade, at least in emerging states that do not have modern governments run by economists. Our historians talk with approval of the "merchant adventurers," the people who sought a broader way of life through trade.

For trade to work, there must be a market for imported commodities, but this market will result from the very increase in population which drives the better-off to trade. The way must always be open, therefore, for sons of the wealthy to find lives of freedom in importing objects that the masses want. We expect trade to develop not in the service of the hungry poor, but in the service of the aspiring middle class. The ecological hypothesis predicts trade to be important in a state only when there are too many people trained to better-class ways. But trade must also have an immediate *effect* on the opportunities open to all classes, because the parent society has to make the objects to be spent in trade.

In creating niche-space (jobs) for children of the wealthy, trade must also create jobs (niche-space) in the parent state. Because people must make things to sell outside, trade multiplies the niche-spaces available in the crowding state. First people can find a broad-niche life by engaging in trade, then other niche-spaces are made at home for those who supply the articles of trade. But even the stay-at-homes are getting part of their living from other people's land. It is quite wrong to think of those who stay at home as being supported by the homeland, because many dimensions of their niches are supplied by the foreign states who take their manufactures.

After trade becomes commonplace, the hypothesis predicts a second and inevitable consequence: the mass of the people will become dependent on imports for their very subsistence, very likely even for their food.

86

They do this because their numbers go on rising *after* trade has become important to the state, as well as before. This late-arriving portion of the population is dependent on imports for necessities. Once, therefore, a state begins seriously to trade, the rising numbers that trade makes possible become dependent on continuing the trade.

This analysis departs drastically from conventional wisdom about trade. We usually think of trading states, say modern England or Japan, as being driven to trade in order to feed their people. Modern politicians in those countries make speeches about "having to export in order to live," which leads people to think that the dense populations came first, and that some desperate necessity drove the crowded masses to resort to foreign commerce. But the ecological analysis denies this. The crowded masses are not a *cause* of trade, but a *consequence* of it. The only way in which crowding causes trade is through the pressure on the lives of the better-off. Children of wealthy people engage in adventurous commerce to maintain their own standards of life. By doing so, they make it possible for more people to be raised in subsequent generations. These new people are the ones who are physically crowded. They are dependent on commerce, certainly, but they only appeared as a result of the commerce started by others.

So trade provides additional broad niches for the traders themselves, niches for the stay-at-homes who supply the goods, niches for those who man the ships, and finally, a large number of narrow, poor niches for the surplus people who are raised as the breeding strategy responds to the extra resources released by trade. And yet another expansion into a new niche comes from trade—that of soldiers.

A trading caravan or ship is a peculiarly attractive object to third parties, embodying as it does concentrated loot. Furthermore, the traders must conduct their essential transactions far from home in an alien state. The trader is both a convenient person to rob and a likely person to blame when any little tension is felt in the host country. Traders, therefore, must be prepared to defend themselves. The parent society must both police the trade routes and be able to mount punitive expeditions.

This need for police power means that skill in war is essential to the trading state. The soldierly professions are added to the niche-space, thus providing far from narrow lives to another cohort of people. And it is necessary that these new soldiers be equipped with the latest weapons and advanced technique, because trade develops as a middle-class endeavor and is a product of clever people seeking to live well. The military means employed by merchants ought to be efficient, reflecting the skills in industry and commerce that made their class wealthy. Trading states like Venice, Japan, or ancient Carthage, whose desperate fate I tell below, all brought technique into service when they made their weapons; and the celebrated doctrine of the "Fleet in being" with which the British of the Empire patrolled the world sea lanes to keep the "pax Britannica" is the most familiar example of the military acumen needed by a trading state.

A civilized soldier employed by the merchants will be armored, for he fights not out of pleasure but from calculated necessity. Getting hurt is to be avoided. Weapons, tactics and discipline will reflect the organized life of his thriving city. The hypothesis predicts, therefore, that an emerging trading state will develop the

best weapons and armor that their technology can produce; the city will take, as it were, a cost-effective attitude to the arts of war.

We can add to the list of predictions of the ecological
hypothesis

That trade will develop as the niche-space of middle and
upper classes becomes crowded.

That opportunities in manufacture increase as trade grows.

That the population rises and grows denser as a consequence
of trade.

That the trading state acquires advanced weapons and an
army.

When numbers and aspirations for broad niches continue to grow beyond what can be accommodated by
trade, then the only expedient left is outright theft. A
growing city-state will certainly find itself in a world
peopled by others less citified and less densely populated. Very likely much land will be full of wandering
herdspeople or nomadic farmers, ways of life that cannot support dense populations. It may even happen that
citified people will find lands still occupied by hunter-
gatherers, as when Europe first thrust itself into North
America. More often the surrounding lands will be in-
habited by people whose ways of life the city folk had
left behind them some generations before. What is certain, however, is that the neighboring land is, by the
standards of the city, underused and undersettled. The
city-state will be surrounded by cultures whose technology of extracting niche-space from the land is inferior
to its own. Taking over this land is an obvious thing to
do.

The ecological hypothesis predicts, therefore, that a society will engage in land theft when its organization and aspirations show it to be better able to extract a living from the surrounding lands than the people already there. And it follows that a society which has reached this position also has, and knows it has, the better weapons.

Land theft means planting a colony or annexing a whole territory. Both processes must be resisted by the people already there. But those of the civilization who covet the land have an advanced technology of war, inevitably, by reason of the technology and trade which has made them a city. When they begin armed emigrations, whether for colonies or for empire, they cannot be stopped. This is why nations like Rome or Greece built such glittering victories.

True colonies represent the simplest form of land theft. You send out soldiers, occupy a piece of land and fill it with settlers. You carry on your own expanded way of life away from the parent city, not so much relieving congestion at home as providing the necessary opportunity for the increased numbers in each generation. When you have many colonies, you might fill in the gaps between and make a small nation-state. All colonization is aggression, but there is a gradient from making a small settlement to wholesale annexation of aggressive war. You use your superior weapons to take niche-space from others, by force.

Aggressions that come about in this way are deliberate affairs executed by superior weaponry. We do not have to think that the aggressor fights like a savage beast. Soldiers annexing land are not fighting in the way an animal fights, and we can safely reject some fashionable ideas about animal passions in human aggression. Passion has nothing at all to do with the matter. The colonizing power does not set upon its neighbors in some paroxysm of fighting hate. Nor are its soldiers spoiling for a fight, triggered by some releaser of rage. The state is calculating. The soldiers are armored and cautious. The enemy is weak and a victim. The object is loot.

When the leading classes of a state have led their people through the stages of technical improvement in manufacture, a class hierarchy, trade and the colonial expropriation of land, they are coming to the end of the possibilities for finding more niche-spaces. Yet a society putting all these into effect is likely to be a buoyant one and its people are likely to be conditioned to the long success story. The breeding strategy, therefore, will certainly work to keep families relatively large. Each couple of the colonial state will choose its family in some hope, and this will be so in both parent city and daughter colony. This means that the succeeding generations will see more people still, not just starving poor but more particularly aspiring upper castes and classes.

All that can now be done by the rulers to keep control is more of what has gone before, and this we must expect: more attempts at trade, more social ranking, more aggression. Aggression seems the most promising alternative.

Sending out a civilized army to take yet more undeveloped land is not only likely to succeed, but also exciting. And so niche theory suggests that a tide of aggression ought to flow out of the expanding state until a time comes when something stops the flood of armies; perhaps the distance of communications, perhaps reaching a boundary defended by some other army of almost comparable technique, perhaps a combination of both.

Aggression remains available as a solution to crowding in the more desirable niches only for as long as the weapons of the state are superior to the weapons of any people within reach. The aggressor state will always be both wealthy and wanting more wealth. Victory will always be achieved through superior technique.

The battle fought by Alexander the Great at Arbela was part of the climactic aggression of the Greek peoples in this way. That is the ecological verdict on the Macedonian triumph. The series of Greek wars, and Alexander's quest for empire, were predictable. A superior civilization had won more resources from traditional lands; it had promoted superior technique, an expectation of continual improvements in the standards of life, and an increasing population, all at the same time. There had long been emigrations as people sought scope for their superior civilization elsewhere, but still the pressures grew on those who stayed at home. The best efforts to improve the ways of life of the people were frustrated, and even the traditional way of life of those in authority became threatened. Petty wars between smaller city-states reflected the general frustration, then more ambitious wars became possible as the civilized peoples invented more effective ways of fighting. Ambitious captains could promise to fulfill the aspirations of their followers with the plunder and land of conquest. Eventually, the chances of battle gave the Greek empire to Philip of Macedon, and his son, Alexander, used this power to overcome all the states within reach of his unbeatable armies.

We will find a similar pattern of events behind all the greater conquests of history. Aggressive conquest is to be expected whenever population and aspirations grow together. Up to now every advance of civilization has been accompanied by rising desires and rising numbers.

Always this has resulted in aggressive war. Ecology's second social law may be written *"Aggressive war is caused by the continued growth of population in a relatively rich society."*

The combined expedients of better government, better technique, emigration and going to war can, of course, never produce more than a temporary relief from the pressures of demand. If numbers go on rising, the condition of the people, both leaders and led, will be as constrained as ever within a few generations at most. However large the empire built from underdeveloped lands, there has always been a finite limit set by logistics and geography. When all is full there is nowhere else to go.

Niche theory predicts, therefore, that a limit will be reached to the number of broader niches that can be found by ingenuity, trade and theft. And yet, the theory also predicts that the numbers desiring broad niches will continue to increase. The empire will become crowded *for its upper classes.* It is this phenomenon which is likely to be the cause of decay. Social unrest is now inevitable.

As the empire crowds, freedom of choice must be an early casualty. There has to be more government to allocate and control. Bureaucracy will be getting more complex, its practitioners more numerous. This is so inevitable a consequence of expansion that a minor ecological social law might be written, *"All expansion causes bureaucracy."*

But the bureaucrats cannot make more resources, they can only allocate what they have. Opportunity for betterment wanes, and initiative must wane with it. The army is no longer the pathway to a good life and will be neglected. After all, the only role soldiers have left is defense against distant barbarians. Once a fresh military power appears at the borders the empire must fall.

93

In the final days the empire may linger on if it can impose so stern a caste system that many families are held small by want. This is what some of the longer-lasting civilizations such as Byzantium and India achieved. But this works only until other states catch up with the static weaponry of the moribund empire. Then comes destruction.

The final set of predictions can be summarized, then, as:

Superior weapons will be used to expropriate land and to plant colonies.

All aggressive enterprises are undertaken with superior military technique and in a calculated manner.

Aggressive wars are launched by rich societies and come from the needs of the comparatively wealthy, not of the poor.

An elaborate bureaucracy and loss of freedom will always appear some generations after the establishment of empire by conquest.

Collapsing empires will have rigid caste hierarchies and stagnant military techniques.

Ecological theory, therefore, predicts a number of plausible events during the growth of human societies. People once lived everywhere in similar and unchanging niches; but, unlike all other animals, they maintained their niches from generation to generation largely through learning. This gave us a global distribution achieved by no other animal. At the end of the ice age our ability to learn from experience set us to changing the niche in ways that released larger, and more certain, supplies of food. Our unchanged breeding strategy then produced patchy but dense populations. Developments beyond this stage involve the changing fortunes of human societies of which historians write. It is, therefore, possible to test these predictions of the theory against the narratives of historians.

In the early time of learning new ways many ancient taboos and feuds against neighbors are seen to be ridiculous and irrelevant: they are done away with. Divisions of labor bring large niches and small; wealth and poverty. But change is in the air, and success at rearing children is assured. The population grows. More people begin to live ample lives, but the poor get more numerous as more people are born than can be rescued from poverty. A city-state densely filled with people, results. This state will seem most crowded to those who are better off, not to the poor for whom crowding makes little difference to daily life. An animal example illustrates the point. Wolves live at very low density, for several square miles of typical wolf habitat are needed to sustain each wolf. But deer can live at ten or a hundred times this density, because a square mile of deer range feeds many deer, though it is not enough to support a single wolf. Wolves are crowded at a few per square mile, because a wolf niche is broad, but deer can be a hundred times as numerous before they feel crowding. So it is with people: the wealthy are the wolves, with broad niches easily hurt by crowding; the poor live densely like the deer and can absorb large numbers before they feel real constraints from crowding. Always in history the wealthy are the first to react to rising numbers. Politicians who imagine that rapidly growing population is a problem mainly for the poor are in grievous error. The lives of the wealthy will always be touched by rising numbers long before the population pressure reaches those whose lives are narrow already. Great historical events are caused by the hemmed-in middle class as the numbers grow.

Adventurous commerce with other lands will be pressed in the interests of the better off. This will be followed by emigration, again of those whose aspirations are largest, and colonies will be set up abroad. The material wealth of the state will be greatly expanded by these means, and more people will live ample lives than ever before. But more people will live in poverty too, even though the actual level of poverty may not get much worse. Colonies and trade provoke both friction with neighbors and the invention of superior techniques of war. The expanded living space is held by force of arms, but there can be no stability because the numbers of its citizens must remorselessly increase.

There must follow more fighting, more colonies, more trade, more friction, more troubles with a proletariat which yet must be held down. The soldierly professions thrive, for only they can keep the peace. Soldiering offers a good life to the middle class, and soldiers are inventive. Aggressive war, waged by a great captain, then leads to the establishment of empire.

The cake of resources is then the largest possible; people feel powerful and free with their ancient way of life assured. But part of that way of life is the belief that the family size is the concern only of the individual family. Every couple continues to have the number of children which it thinks it can afford and the population continues to rise. As the resources can be increased no further, so the numbers who must be deprived increase also.

Freedom wanes within the empire as the people in charge seek to maintain ancient standards through more government. Bureaucrats allocate; social divides harden; the possibilities for adventure decline. The ruling classes, once at least sprinkled with people who talked of preserving and extending the people's liberties, become self-centered, repressive, concerned only with the preservation of their own privileges. War is no longer a way to social and material success and will become an unpopular profession.

The mass, once ready enough to take cheer in the successes of the state, must lose hope of better things and may be appeased only by a system of welfare. They know very well that their lot will not improve, that their children have no chance to live better than they, for the evidence of resources pressed to the limit will be visible on every hand. They get on with the business of living as best they may, raising as many children as they can afford until the pressure of numbers and the increasing unhappiness of life sap the spirit of cooperation that is the fabric of a state.

Since interest in war has long been muted, the empire's military technique stagnates and it may well be destroyed by exuberant soldiers from outside, bent on loot and land. Until the foreign armies come, the state may persist for a long time as a caste-ridden bureaucracy, perhaps with a thinly settled countryside, certainly with a dense urban poor. All this seems directly predictable from an ecological hypothesis of a human animal that learned to change its niche to increase its resources almost at will, but without changing the animal breeding strategy of raising the largest number of offspring for which any couple could provide.

In *A Study of History* Arnold Toynbee traces the rise and fall of all civilizations of which we have record, find-

97

ing a common pattern. The twelve volumes of his book occupy more than a foot and a half of shelf space in my library. The supply of historical facts is copious, and the accuracy of the reporting apparently hard to reproach. Yet Toynbee develops ideas about the spiritual drives of human societies that now make the work seem old in its thinking. Toynbee sees civilizations growing at the service of some mystic force, with religion as almost a guiding purpose. Such a work may seem a poor place to look for evidence of a Darwinian interpretation of history based on breeding strategy and niche. But we can take Toynbee's facts without following his mystic interpretations. *A Study of History* is a comparative study of twenty-one civilizations. Toynbee shows that the fates of all twenty-one were alike, stage by stage, as they rose and fell.

Civilizations arise in marginal lands; Toynbee says that people need the spiritual shock of a hard environment to give of their best. An ecologist is not surprised to learn that marginal lands foster aggressive civilizations, though less impressed by Toynbee's belief that the arousing of the spirit is what counts. It is in marginal lands that the pressure of rising numbers will be felt first, forcing expansionist zeal. The limits are reached sooner in marginal lands, habits must be changed sooner if want is to be avoided, and aggressive armies become an earlier requirement. Once the armies are made, the people of a marginal land only need a victim for their aggression, and a plump victim is always by definition waiting next door.

From then on, Toynbee's reconstruction is as predicted by the ecological hypothesis. There is, for a time, a "creative minority" of people whose example is willingly followed by the mass, but the "creative minority" slowly changes to a repressive "dominant minority." The mass no longer emulates, becoming instead a sullen "internal proletariat." The change is gradual, the shades of the proletarian condition are many, but the direction of the changes always obvious. They are accompanied by continued and protracted warfare, a "time of troubles," a time when people feud and intrigue against each other, when plunder and petty kingship seem to be the goals of the most vigorous citizens. But eventually, often after several generations of turmoil, a more able chieftain than the rest imposes his military will; the people gather thankfully behind the prospect of peace which he offers, yielding to him the instrument with which to establish an empire. The evolving civilization has culminated in a "universal state," and there may follow a protracted period, as under the Roman Empire, when an ordered society persists, the dominant minority remaining in charge, the mass constituting the "internal proletariat" consenting or collaborating in its bondage. But in the end the social order always decays. The "dominant minority" is hard put to defend the boundaries of the empire against neighboring peoples and these evolve into a hostile alien force, the "external proletariat." Finally, with the breakdown of order within, and the increasing hostility of the less-disciplined but freer spirits from without, the empire crumbles leaving behind only traces of its culture and religion out of which those who have inherited its impoverished land can begin again the process of invention and order.

Those were Toynbee's conclusions on the rise and fall of all the twenty-one major civilizations that he thought he could recognize from a long perusal of history. The creative minority, the successor dominant minority, the time of troubles which ushers in the succession, the internal proletariat, the emerging conqueror or great captain whom Toynbee sometimes calls a "saviour with a sword," the universal state that he builds, and the long autumn of order while the state endures as a stable thing with all knowing their place—are all events predicted by the ecological hypothesis.

In the mechanisms of the times of trouble, which forge the internal proletariat and pave the way for the great captain and his armies, can be found some of the more revealing workings of ecological process. Not only do small crowded states war with each other, but capitalist business always emerges with all the social problems it brings in train. This is society trying to increase resources with improved technique, working to increase the size of the cake at home while its armies are fighting to increase it abroad.

From the time of troubles onward the population shifts within the state. There is a drift from the land as peasants are displaced in the interests of increased production. This happened in ancient Greece and Rome no less than in the time of the enclosures in Tudor England, or in modern industrial states. The process can be seen in the development of every civilization. Feeding great numbers of people is more easily done if they are brought together in dense settlements; running the agriculture needed to supply those dense settlements is more efficient in the larger agricultural units of agribusiness. And the "drift from the land" is a predictable consequence of a dense population growing denser. It follows, therefore, that partial depopulation of the countryside results from population growth: the land looses people even as the total population climbs. It is easy to mistake this loss of people from the country districts as evidence of a population fall. Roman Pliny made this mistake and there are historians who have followed his errors to this day. But emptying of country districts must be a usual consequence of a population rise, not of a fall. The modern United States of America is an example.

In actual human practice, the first-order reason for driving peasants from the land was to enrich the landlords, but societies put up with the miserable injustice involved because the new ways were more productive; letting the landlords have their way yielded more food for the state, and if pushing even more people into the growing populations of the towns' displaced peasants, it at least promised bread for those growing populations. This is the argument of our "green revolution," an argument that has been used as long as there have been civilized states. "The people must be fed; farm the land efficiently for the benefit of dense settlements."

The surplus people of the countryside go to swell the ranks of Toynbee's "internal proletariat," already being bred in the cities. To these are added the inhabitants of conquered less-developed lands, driven in turn from their fields by civilized businessmen. Spent soldiers join them, their stipulated service over and their military usefulness gone. So there always developed in the great cities of empires large and growing populations with very little to do except be house servants or formal slaves of that dominant minority which expanded to provide both the governors and the bureaucrats of the state. All could be given bread for a long time by improving the efficiency of agriculture, and by taking the new efficient method to the freshly conquered lands of the spreading empire. But each new advance of technology, or regiment, sends its own quota of exiles to the central city, there to join that breeding proletariat in, but not of, the culture of the times.

So far Toynbee's history goes as an ecologist would expect. Endless growing numbers both maintain poverty and give it such institutional form as slavery. The new technology which increases resources is never able to exceed the demands made upon it by ever more mouths, and its only real result is the closer herding of people, together with the growing bureaucracy needed to constrain them. Wars of conquest relieve matters, but only for a short time. When the victory has been truly great, then the numbers who can aspire and live in large niches is expanded for a while so that hope can flourish also. This is why conquering societies talk so much of freedom and liberty. But the increased living space must be filled quickly by the broad-niche "species" in society so that the hopes of succeeding generations must be curbed. Then the needs of an ever-growing proletariat must press upon even the large living space won by conquest. The only stability then is the short-lived one of people knowing their place.

But an empire always has an edge, at first diffuse and spreading, but later almost stationary. Its actual position is a function of contemporary technique in warfare, government and transport. Outside it live people whose kindred have been conquered, absorbed, oppressed, deported, or even slaughtered by soldiers of the civilized state. The survivors beyond the pale have learned something of the civilization's technique in fighting, and have also learned to lead mobile lives, so that they may avoid forays of the empire's soldiers. They take revenge for past injustice, when they can, with raids on the empire's granaries. A whole new way of life, a new niche, has thus been developed in response to the pressure of the

empire's people. Footloose, self-sufficient, partly nomadic, warlike; it is a way of life which often seems
admirable to the imperial governors who confront it.
Tacitus wrote admiringly of the manners and virtue of
the German tribes, comparing them with the degener-
ate society which he saw in Rome. And many a British
soldier murmured the lines of Kipling extolling the
courage of the people of Afghanistan. Toynbee finds
such people living on the boundaries of all the civilizations he studied, calls them the "external proletariat,"
admires them, and tells us that they have in common
the writing of epic poetry.

In Toynbee's account the war bands of the external
proletariat are eventually to cut their way into the dying
empire, hastening its fall. They are those same barbarians outside, who Gibbon tells us tore down a Rome
weakened from within by free bread, circuses and civil
war. Among their numbers are the thrilling names of
Goths, Franks, English, Lombards, Huns—names made
immortal in the Roman Empire's successor states.

An external proletariat, forged from people who do
not care to live in cities but who must run from a civilization's soldiers, is, with the knowledge of hindsight, a
plausible outcome of the building of an empire. But it is
not something which could be readily predicted from
ecological principles alone; it is an outcome of the social
habits of the human species, and you cannot predict it
from knowing that people change their niches while
persisting in raising all the children those new niches let
them afford. What the ecological hypothesis can predict, however, is that once the external proletariat has
perfected the new way of life its own numbers will tend
to rise, forcing it always to look for more resources both
to safeguard children from want and to apportion

among its younger sons. Trained to war by their way of life and equipped with some of the military techniques borrowed from the empire, their obvious expediency becomes armed raids across the borders. Their pressures on the empire must grow as their numbers rise, and an empire steadily weakened by the pressure of its own miserable masses finds itself ever more strongly attacked by armed young men from outside.

The fading summer of each of Toynbee's civilizations passes with the muted mutter of dissension in the big cities and ceaseless petty war at the frontier. But, at last, government crumbles at home, and war bands from outside swell over the disintegrating mass. Toynbee draws lessons of the spirit from these events, looking for the prime causes in the class war, the failing vigor of a privileged minority and the social injustice of commercial exploitation at home, contrasted with the ennobling experience of the external proletariat, which left it independent and tough. Moral virtue then triumphs over moral decay and the lands of the empire are inherited by new peoples who proceed to build a new civilization on its ruins. No doubt these events do strange things to the human spirit, but their prime cause is not spiritual; it is an animal breeding strategy applied to human affairs. All of what Toynbee sees is predicted by the ecological hypothesis; all is the inevitable consequence of trying to provide a better life for ever increasing numbers of people.

In the end Toynbee notes that a world religion rises from the oppressed proletariat and persists long after the empire has fallen. He claims that all the major religions of the world arose in this way; they started as religions of those subjugated in empires. Ours of the West was one, built out of conquered people desperate under the exactions of Roman military rule. Ecologists can easily understand the form these religions take. Much of the appeal of proletarian world religions lies in their counsel to the oppressed to endure. Nothing can be done; the poor are with us always; rely on your spiritual strength and make the best of things. For the crowded masses, to whom Jesus, Buddha, Confucius, and their like appealed, there was no hope for an improvement in the standard of life. People knew in their bones that the lives of their children would be no better than their own lives. They did not know that the reason for this was the swelling numbers of people who used up new resources as fast as they could be created, but they truly knew the outcome all the same. So, by the device of a new religion, crowded, poor, citified people have always turned to the plastic properties of the human spirit, learning to be happy with very little. People learn to live in very narrow niches when religion teaches, and world religion is but another expression of learning the necessities of life by the process of taboo.

CHAPTER FIVE

WAR IN THE MEDITERRANEAN

TWO THOUSAND four hundred and seventy years ago a line of armored spearmen ran across the beaches near the little Greek village of Marathon, cut aside the wicker shields of a Persian army, and stabbed the invaders back into the sea. The armored spearmen represented a new technique of warfare, which was to be slowly perfected into the terrible Macedonian Phalanx with which Alexander conquered 160 years later. The Persian soldiers were not equipped to meet the shock of an attack by an armored formation. They had little defensive armor themselves and equally little chance of beating through the enemy armor at close quarters. The Persians did have many bowmen, who served well enough against traditional Persian enemies, and it was through a cloud of spattering arrows that the armored lines of Greeks advanced. The Greek soldiers came through the arrow cloud at the run, despite their heavy armor, keeping the time when they were shot at to a minimum. And when the panting lines had trotted through the short distance that bowmen could sweep it was all over with the Persians. Flesh met steel and victory was given to the better technique.

The Persian army was but a minor force sent out from a very large empire to punish impertinent city states at the border. The Persian ruling class can have had no idea of the oblivion that those armored spearmen were to bring to their powerful state. The Greek soldiers, jubilant in what must have seemed a David-against-Goliath success, cannot have seen what they then had the power to do. But the battle of Marathon was a signal that the history of the modern West had begun. A civilization with its own advanced technologies had worked out its ways of living. The niches of the people in this new civilization included so much trade and colonial enterprise that needed fighting that they had forged an instrument of war that could overwhelm traditional contemporary armies. It is from the time of the battle of Marathon that we trace the hegemony of European power.

The people living round the Mediterranean Sea were to develop a common culture, shielded from the massive Oriental despotisms to the east by superior weapons and training for war. This Mediterranean culture was to reveal all those changes in the human niche that are to be expected if the people's hopes rise even as their numbers crowd. There was to be triumph, freedom and wealth; but also poverty, hopelessness, military neglect and collapse. People were to respond to the changing circumstances of their times by forging new religions appropriate to the niches in which they had to live. The Mediterranean Sea became a theater for an ecological play, as the people living round it learned better ways of life even as their numbers grew. The curtain rises on this play at Marathon.

Before the Greek armies appeared the centers of learning, cultivated living and military power were Oriental. The Persian empire was organized out of ancient civilizations, all of which had been centered in the Middle East and Asia. Asiatic peoples had lived in empires for two thousand years and more by the time of Marathon, populous empires of ingenious people, technically proficient in agriculture and the engineering of cities, governed by remote sovereigns who worked through elaborate bureaucracies. There had been nothing to compare with this in Europe, most of which, from the hinterland of Greece to Spain, was still barbaric. There were cities of some antiquity on the coasts of Greece, its islands and Italy, but none had debouched into empire. The Persians still talked of Greeks as "barbarians," though perhaps from pride rather than accuracy.

But by the time of Marathon powerful states were beginning to appear on the Mediterranean coasts, each with soldiers or fleets well provided with civilized technique and training: Greece with its phalanx of spearmen; Carthage with its fighting sailors; Rome with its military mystique. Each expanded, each debated freedom and made constitutions, each went to war. Greece held the Persians, threw them back, found unity in battle and at last, under Alexander, overwhelmed a Persian empire, already crumbling from within and no longer a first-line fighting power.

Farther west a military republic in Rome fought its way to the government of Italy, and then quarreled with Carthage across the water over trading rights and who should hold Sicily. Rome and Carthage were isolated from Asiatic imperial power by the facts of geography and they were able to expand populations and hopes against each other, with some common friction against Greece, but otherwise in a one-to-one collision. And so were fought three terrible and celebrated wars. Fleets of triremes, rowed warships equipped with underwater rams and missile-throwing machines, clashed over the lengths of the Mediterranean Sea and Hannibal crossed the Alps with his elephants to fight his way almost to the gates of the Roman city. These Punic Wars afford some of the nicest illustrations of how people may be drawn into inexorable combat by ecological need and they are probably the best guide we have to the likely cause and course of future wars which start with nuclear attack. They resulted in the annihilation of Carthage and the appropriation of her resources to the victor, Rome.

In these fights, first with the states of Italy and then with Carthage, the Romans invented their own new instrument of war. They first perfected a phalanx along the Greek lines, but they did not stop with the phalanx. They went on to invent the technical answer to the phalanx, a wholly new formation and a discipline for war which we know as the legion. This legion let the Romans do what the Persians had failed to do—smash the armored lines of Greek spearmen and conquer all of Greece. The Romans went on to subdue the remnants of Alexander's Asiatic conquests, and to absorb the barbarian regions of the rest of surrounding Europe which had as yet developed no advanced city-states of their own. The Roman Empire had come.

Then followed the familiar cycle: republic to civil war to war lords to empire to imposed peace. The Roman peace of the *pax Romanum* faltered into new civil war, dictators and despots imposed, and reimposed, order but there were such discontents as even ruthless government could not still, the army failed to hold frontier line after frontier line until armed barbarians fought their way in to loot the cities and settle the land. The Roman economic, technical and military system disappeared and new ones slowly took their place.

The Christian feudal states of Europe emerged as the barbarian conquerors learned ways of life suited to the old imperial lands they had won. In North Africa peoples from the fringe of the old Oriental despotisms expelled the penetrating barbarians of Europe from their side of the Mediterranean and built a different culture out of the imperial remnant, the culture of Islam. The Christian and Muslim cultures then split the Roman imperial lands between them, one to the north the other to the south. All this was decided 1,250 years ago, after Charles the Hammer fended a Muslim army from France at the battle of Tours, and King Leo of Constantinople held his fortress against Islamic armies reaching into Europe from the East. After this the Mediterranean was no longer a theater on its own, and the pressures of expanding peoples henceforth radiated from northern Europe, eventually finding their temporary relief in the scarcely populated lands of America and elsewhere.

Such, in barest outline, is a story of human expansion which an ecologist might fairly call an episode, the "Mediterranean Episode." Its time was short, merely 1,212 years from start to finish, only forty or fifty generations, a time too short for much significant change in niche to happen by genetic selection, but long enough for numbers to change almost beyond imagining as people changed the niche by nongenetic means as only people can.

Greek, Carthaginian and Roman civilization came out of city states built by settled tribes of people who were once barbarian. As the city states grew, they warred with neighbors who were still barbaric. The growing Mediterranean empires fought continually with barbarians at their frontiers and were fated, at the last, to lose their imperial spaces to barbaric armies. From barbarians they came; before barbarians they fell; in the qualities of barbaric life, therefore, is civilized history based.

We use the words "barbaric" and "barbarian" as if they referred to people who were necessarily violent in their ways; people who were crude, unmannerly, coarse, unappreciative of beauty, or who were actually brutal and base. But this is not what the ancient writers meant by "barbarian" nor are these habits typical of barbarian peoples of any time.

Barbarians are simply people who have not got used to living in cities, and who have not adopted the economic structures which city life requires. The essential quality of barbaric life is living off what the immediate neighborhood provides. Barbarians are farmers, or herdspeople, living in family units or tribes of closely related people which are self-sufficient. They are not used to the refinements and luxuries of city life, and their manners may seem strange to those who have been molded and constrained to the ways which a city demands. It is this which gives a barbarian an aura of wildness to cityfied people. But barbarians have always had their own subtle codes of manners, suited to their country and unsettled ways of life, yet no less gentle than those different manners required by a city. If anyone doubts this let them arrange an invitation, for instance, to the hospitality of an eskimo's home, where manners are gracious and so obviously suited to the way of life.

In the arts, no less than in manners, the barbarian peoples of the ancient Mediterranean world, like their

111

successors down to the present day, have woven beautiful things into their daily lives. The ancient writers have not bothered to tell us of the beautiful things that *their* barbarians may have made, but archaeology has sometimes done it for us instead. Recent excavations in the frozen soil of central Asia have shown us, in the form of tapestries and paintings, of what loveliness nomadic barbarians were capable. More in our everyday experience is the beauty of Oriental carpets, many of them typically the product of peoples living in ways which an ancient classical writer would have called "barbaric." The Turkoman tribes of Central Asia use their elegant pile weaving to make not just floor coverings, but saddle bags, grain sacks, flaps to the doorways of their tents, and even the broad bands, or lashings, that hold the tents together. It is "barbaric" to work long and hard to make a beautiful bag in which to keep a family's store of grain.

The tribal peoples of Afghanistan may give us some idea of the barbarians from whom the Greeks came, and who eventually cut their way into the Roman Empire. The Afghans still make very lovely carpets, to my thinking preferable to the more elaborate designs made in the cities of modern Persia. And yet the Afghan communities who make those carpets are, many of them, nomadic peasants, living from flocks and herds. Not so long ago they waged a stubborn and triumphant struggle against the civilized soldiers of a British empire at whose border they had the misfortune to live.

There is no doubt at all that the way of life of a barbarian can support far fewer people in any country than can civilization. Barbaric agriculture cannot be intensive; the immediate land must be used to provide all the necessities for life, those for which it is not well suited as well as those for which it is perfect. The community grows its own corn, raises its own meat, finds its own herbs, and produces the raw materials for its own clothing. Although scarce items can be traded over long distances, even among nomadic peoples, the difficulties of transport and the bulk of staple commodities keep trade down to a minimum. Furthermore, the barbaric way of life finds its emotional satisfactions in the use of space. The whole of barbarian behavior and code of manners, which civilized people have affected to despise, is attuned to living with elbowroom. There is a bit of the nomad in all barbarians and some, of course, adopt an almost completely nomadic way of life. So, both resources in the simple economic sense of food and raw materials, and resources in the ecological sense of all the requirements of the adopted niche are such that populations of barbarians cannot be very large. They feel crowded when their population density is yet low.

A barbarian way of life is also susceptible to vagaries of the seasons and weather, so that the tranquillity of living must be conditioned to interruptions from those acts of nature which the insurance industry describes as "Acts of God." For barbarians there are lucky generations and unlucky; sometimes the numbers rise, but at other times privations make them fall. People are inclined to move if there are too many of them, and they are also inclined to move in response to local disaster. But moving will always bring you to land occupied by other barbarians who may be hard-pressed enough to live, without accommodating newcomers. Barbarians must, therefore, often be prepared to fight for the lands and resources which make their way of life possible. This accounts for the warlike qualities of barbarians

which have made them both feared and disliked by civilized writers.

But though barbarian peoples may be warlike in the
sense of being both vigorous and brave, their way of life
usually sets limits to both the technology and the discipline that they can bring to war. Individual bravery,
initiative rather than conformity, varied and primitive
weapons; these are the qualities usually found in barbarian fighting forces. Civilized states can usually
triumph against even the most warlike barbarians,
through superior weapons, standardized techniques,
and the reinforcement of courage with training and
obedience.

The immediate effect of switching, even gradually,
from barbarism to a settled city life is that the population grows. The new economy produces more food, the
new niche permits people to be content when living
more closely packed, each couple can raise more children and does so, and the numbers of the people steadily increase. So it was in Greece.

The ecological hypothesis predicts that this process
will lead to colonial enterprise, to trade, to much fighting, to an oppressed proletariat, to high technology in
war, and to the creation of empire by military means as
a popular goal. The written history of Greece shows
how each of these things came about, and even reveals
that the Greeks themselves knew something of what was
happening to them.

The pattern of these various consequences of rising
numbers and changing niche in Greece was influenced
strongly by Greek geography. The land is both moun-
tainous and dissected by embayments of the sea. This
meant that scattered city states could grow in a partial
isolation from each other, having defensible state
boundaries and well limited patches of local resource. It

may well have been this isolation of each Greek city-state that helped to foster the refinement of their remarkable military hardware, protracting wars between neighbors, letting victories be indecisive so that return engagements could follow after a few years spent in perfecting armament. Certainly it is true that the first and most fundamental of civilized fighting forms, the armored phalanx, was better refined in Greece than in any nation of which we have record.

Standing in line, fighting with spears, and huddling behind a row of shields is the simplest of organized military forms. We can trace these essentials of the phalanx in the early days of every civilization so that the Egyptian and Asiatic powers knew of these things long before the Greeks emerged from barbarism. But the Greeks brought to the phalanx both armor and training on a scale that was wholly novel. Each first-line Greek soldier (*hoplite,* as he was called) had a load of steel and other metals encasing his body which weighed more than seventy pounds. It was this great weight of helmet, breastplate, thigh pieces, greaves and shield which made the running advance at Marathon so remarkable an exploit. Thus equipped, and standing close by his neighbors so that their shields made an unbroken wall, Greek fighting men would be safe enough if their enemies were mobs of sword-waving barbarians, however brave. Even the soldiers of imperial Persia were to find that they could do very little against these lines of armor. Greek observers of the Persian wars saw this clearly enough, as this quotation from Herodotus shows.

The Persians were not a whit inferior to the Greeks; but they were without bucklers, untrained, and far below the enemy in respect of skill in arms. Sometimes singly, sometimes in bodies of ten, now fewer and now more in number, they dashed forward upon the Spartan ranks, and so perished. . . . Their light clothing, and want of bucklers, were of the greatest hurt to them: for they had to contend against men heavily armored, while they themselves were without any such defense.

HERODOTUS, *On the Battle of Plataea*

The Greek armies of these earlier Persian wars had not yet reached that ultimate refinement of Macedon with which Alexander was to conquer, but they were already dreadfully effective against any force that was not Greek-equipped and Greek-manned. An echo of this comes from that most moving of all the stories of ancient fights, the defense of the pass at Thermopylae by Leonidas and his three hundred.

A massive Persian army was marching southward down the Greek coast, supplied and nurtured by Persian fleets in an amphibious operation brilliantly conceived and executed by the Persian general staff. The Persians had refused to let their invading regiments be sucked into the mountains of interior Greece, where they could not be succored by the supply ships of the fleet, and were moving to cut off and secure the Greek maritime cities. Their strategy can be likened to that of Eisenhower's armies after D day as, supported from the sea, they fought their way along the coast to secure the port of Cherbourg. But the route the Persians followed was a narrow route, pressed between the coastal mountains and the sea. There was a coast road, perhaps half a mile of level ground, and then the rising shoulders of the hill. A small army under Leonidas was rushed to hold this coastal pass while the Greek cities could manage the politics necessary to bring together their main fleets and regiments in the common cause.

Leonidas had, at first, less than two thousand soldiers for the defense. Pack this number shoulder to shoulder, line them up several ranks deep, and see how slight a barrier you have built to block a coast road and the open fields by the sea. Then bring against them ten times as many regiments, professional soldiers of an imperial power, used to victory, knowing that food and safety were rewards for removing the Greek obstacle to reaching your fleet, with its victuals, further down the road. Then you have an idea of what was accomplished at Thermopylae before the final betrayal.

The Persians did not attack at once but waited five days. Probably they were hoping for more contact with the fleet, but then they had already learned what it was to charge Greek armor, and their caution is understandable. But on the fifth day the Persian regiments moved, rank after rank, block after block, massed down the road, trampling to the water's edge, lapping upward to the shoulder of the hill. By now Leonidas had been reinforced and his thin steel line was unbroken from hill to sea. Calm, citizen eyes looked out from beside the nose pieces of helmets, the round shields overlapped in unbroken barricades of steel and brass, the thousands of sharpened spear points were held steady and waiting. On the Persian side the shouting, fury, and dust cloud of frenzied attack; on the Greek side calm waiting, the hoarse shout of command, the disciplined united thrust of spears, a cohort at a time. In such a fight, numbers are no help. Herodotus tells us simply that the Persians could make no headway against the heavy armor of the Greeks. Yes, indeed.

There were days of vain attack on the Greek-shield wall, before treachery came to the Persian aid. A renegade Greek conducted Persian regiments through the hills by night to take the defenders in the rear. Sentries gave the warning in time for the Greeks to retreat, all except Leonidas and his three hundred, who chose to remain. What happened then is best told in the Greek couplet which is said later to have been written on a memorial in the pass, "Tell them in Lacedaemon, passer-by, that here, obedient to thy laws, we lie."

The defense of the pass at Thermopylae served its purpose as the Greek armies massed and the Greek warships converged. An immediate result was the sea victory of Salamis, in which the Persian fleet was hemmed in to a narrow channel and destroyed in detail. Salamis was an admirals' battle rather than the sort of clinical application of superior weapons to which Greek foot soldiers were becoming accustomed, because any ship fighting with rams and boarding parties was much like another. Herodotus, however, gives us a clue that even for naval warfare the Greek designers may have given their fighting men the technical advantage, because he talks of the way in which the "heavier" Greek warships shattered the oars of the Persian galleys.

Highly civilized weaponry was a fact of life to free Greek citizens. Every independent man owned the weapons of hoplite infantry and knew how to use them. His own money equipped him for war. He could be "called to the colors" at any time, and he went willingly. Front line soldiering was both the duty and the privilege of the substantial citizen, and the poorer classes went to war merely in his support. The better-off fought; the less-well-off supported. And this was reasonable, because the colonies and trade on which the high standards of life of the wealthy in a Greek city depended could be guaranteed only with expensive weapons.

Each of the major city-states of Greece sent out colonists to found tributary cities elsewhere; scattered round the coasts of the Persian dominions, in North Africa, in Sicily and beyond. This was how local city states, each pressed into an ancestral valley of the rugged Greek terrain, had found opportunity for businessmen and adventurers alike. Building a colony was a well-understood proceeding, involving warships and soldiers. The soldiers were needed particularly because planting a colony always required replacing the original inhabitants. Would-be colonial powers have always had to face this need for soldiers because all the habitable portions of the earth have been full of people since before the invention of civilization. These people might be barbaric, and in low density, but the land was certainly "full" in the ecological sense of not being able to support any more people living in the established niche. The brute needs of survival always have meant that the civilized settlers will be resisted and attacked. So you send soldiers.

The people of Chalcis lived on the fertile plain of an island close to the Greek mainland. When they needed fresh opportunity they sent people away in ships to find new land, and their mariners looked for nice, level pieces of fertile soil like that of their native island. When likely lands were found, they put on their hoplite armor, formed their wall of spears and shields, and the deadly line then tramped forward across the fields it had come to claim. They probably thought no more of eliminating barbarians from lands they wanted than did the English settlers of America think of removing Indians with powder and ball.

The state of Athens grew through colonies and trade in a somewhat different way. The land was not lastingly fertile, like the island of the Chalcidians, having soils which were easily denuded by agriculture. Ecologists know well the Mediterranean soils, like those of Attica where Athens was built, and the story they have to tell. They are now red, being given the name of "terra rosa." This is the red of minerals weathered under a mild climate. But once, in their forested antiquity, the soils were probably brown, because they were well mixed with the humus and leaf litter of the forest above. Good agricultural soils need such an admixture of humus. But the people of Attica cleared the forests, burned off the brush, plowed the land, took away the crops to eat in their villages, and let the burning sun of the Mediterranean dry the soils so that wind and water could sweep the humus away. The fertile brown color went, and the unproductive red mineral mass of the terra rosa remained. This was the result of the first intensive agriculture in Attica, and the Greeks themselves understood the cause. A sentence in Plato reads, "all the rich, soft soil has molted away, leaving a country of skin and bones."

The people of Athens could find more niche-space for the Athenian way of life by building colonies which would duplicate Athenian ways, and they did so. But the poorness of the soils of Attica seemed to have helped the trading side of colonial life to grow with particular energy. Athenian men of business concentrated on taking from their own land only what it would yield easily, which happened to be olive oil and silver, and they proceeded to build ships so that they could trade these commodities for the other things they needed. There was no living for farmers any more, except for the few who tended the olive trees, because the people's grain was now grown by barbarians in the Ukraine, and the people had to crowd near the granaries in Athens and become the servants of manufacture.

There were now rich and poor in Attica as there had never been before.

Through commerce, and the flood of alien cultures swept into the growing city by the Athenian fleets, the lives of the governors were enriched, and their niche was broadened. But once-free farming folk were drawn into the restricted life of a city proletariat. The riches of many lands were drawn into Attica, constantly increasing the wealth and the resources for a good life, but the population went on growing and was sucked into the city to absorb the new resources; a process with which we are very familiar today. And the Athenians, no less than the Chalcidians, had need of that most vital of Greek inventions, the phalanx of armored spearmen. Their rights to trade abroad must be defended, and the rich resources of Athens could be collected so temptingly in one place only if the military means were available to make them secure.

In addition to colonies and trade as answers to the needs of growing numbers, there is the expedient of direct conquest and elimination of neighbors. In a nation which had invented such a clinically effective instrument of compulsion as the phalanx, this expedient was sure to be tried, and it was. The most celebrated exponent of this art of neighborly aggression was Sparta. But aggression on neighbors meant fighting other Greeks who also knew the effectiveness of lines of armored spearmen. Spartans could not have the technical superiority over fellow Greeks which made it so easy to force other nations to yield for Greek colonies, and had to develop a society organized around the needs for absolute military efficiency in order to prevail.

Spartan discipline is legendary. But it is important to note that it was the well born youths who were trained to this asceticism in war. In battle, the hoplite shield wall held by young men of good Spartan families was supported in the rear by up to eight ranks of *helots* or slaves, who passed forward spare weapons. It was this perfection of armored warfare from Sparta against which Herodotus tells us the Persian waves of infantry broke at Plataea, and it was this philosophy of war which drove Leonidas to make his last stand with the three hundred. All this Spartan excellence in war was clearly and directly the achievement of the ruling class and in its own interest.

Both the Spartan military society and the commercial society of Athens worked by compressing the niches of the mass; specialized labor was needed, often dull, repetitive, mechanical, soulless labor. It must be performed by people whose ancestors, only a few generations back, were free farmers; and in a world where free, farming, barbarian societies still existed on all sides. Freedom beckoned in memory and by example. So the proper functioning of the state required compulsion. The poverty of people with compressed lives, which is always the result of letting populations rise to soak up the resources released by new technology or conquest, took on the special institutionalized form of slavery. A slave was merely a poor man made to keep quiet about his inevitable lot by physical coercion.

The Greek city states had already, by the time of the Persian wars, found themselves in another of the dilemmas of growing numbers and ambition—depopulation of the countryside at the same time that the towns became crowded. This is a normal consequence of growth. As agriculture becomes technically proficient, the land can be worked with fewer people. More importantly, the newer ambitious niches which are being taught by example are hard to learn and adopt from a remote rural base. The desire to learn them is the desire to be civilized, and being civilized is a process that starts in cities. Then again, many of the new opportunities are to be found connected with trade and foreign travel, and these things too start in cities. So people leave the countryside as they grow up as a necessary move in the learning of the new niche.

As people better themselves by the trade and industry of the city, so it often happens that the city comes to support itself on the products of that trade. It may well be that the city even comes to meet its basic needs, as for food, from regions other than its own original hinterland. This will be particularly so if the city begins to trade with fertile agricultural states whose lower standards of living let them sell food to the city cheaply. The result is neglect of the ancestral countryside and an even more rapid drift of the country people into the cities. It is a curious paradox that a growing population in a civilized state may actually depopulate large areas of the original real estate. The paradox has led demographers, modern as well as ancient, into thinking the overall population was falling when in fact it continued to grow. It is this process which is behind a widespread belief, probably erroneous, that populations were falling at times in the later Roman empire.

The Greek historian Thucydides saw the working of this process when he noted that Athens and Corinth were "crowded" and had to feed their people with grain imported from the Ukraine. Notice that he was not using the word "crowded" to connote wretchedness, for he thought of all the well-to-do people of these very great cities as part of the crowd. The people were living on Ukrainian land which fed them, and on all the coasts with which they traded, so they had ample resources, though "crowded." Interestingly, Thucydides did not think Sparta to be "crowded" in this way, though he notes that the Spartans did import their food from Sicily. Sparta's land empire apparently left its people less densely concentrated, though she too was living on the produce of other people's real estate.

So the individual city-states of Greece each grew out of barbarism through settlement, manufacture, trade, colonies and dense concentrations of urban people to an eventual dependence on imported food for the large numbers of their proletariat even as predicted by the ecological hypothesis. The wealthier ranks of their societies took to war as they tried to expand and defend their broad niches, and they invented advanced techniques of fighting, particularly stressing body armor. They formed confederations of cities to meet attacks from the powerful, especially when their expansions into Asia provoked attacks by imperial Persian armies. The hypothesis predicts next that increasing demands made upon each city government would result in such wars that the different parts of the growing nation will come under strong central rule and loose its armies toward imperial conquests of its own. All this was about to happen.

Athens tried for conquest first. The Athenian power was based on ships and commerce, and the Athenian vital interest was to the East. The successful wars with Persia went particularly to the Athenian advantage because they freed Hellenic ports and colonies on the Asiatic coasts within the Athenian sphere of interest. Athens put together a confederacy of allied cities and began to grow by military means. Her achievements in war were outstanding, as her leading citizens spent their

wealth on warships and donned hoplite armor themselves. She fought sea battles with the Greek states led
by Sparta, with Asiatic fleets, and even with North African ships from Carthage at the same time. And she won
them all, diverting trade and loot to Athens and putting
up in triumph the splendid architecture whose traces
we still see.

But the Athenian power was in ships and money
rather than foot soldiers as befitted a merchant city.
Pressed against her was always the threat of the massive
Spartan military machine. In any war of attrition between the deadly Spartan armies and the much smaller
land forces of Athens and her allies there could be but
one end—unless Athenian generals should invent some
technical trick to give them the advantage over the Spartan phalanx, which they never did. So Athens tried the
classic tactic of maritime powers, that of cutting the sea
lanes that supplied her enemies with food. But in this
she failed. The Spartan food supplies came from the
island of Sicily and could be interrupted only if the fortified Sicilian city of Syracuse could be stormed and
its harbor used as a base. The Athenians used up
their great treasure in ships in attempting to storm
Syracuse and, through what are usually described as
a series of blunders, were beaten back, their fleets destroyed.

Athens now faced a stark reality faced by many another trading power since. She had waged war with
money and capital which had been slowly built by trade
and that capital had been spent. A series of defeats, or
of victories too dearly bought, as happened to Britain in
modern times, strips a trading power of wealth that has
been laboriously built out of resources in faraway lands.
The Athenian savings were gone, it was hard to buy

more ships or to hire soldiers, whereas the territorial expanse of Sparta was intact, yielding money and soldiers as before. It was not long before work crews instructed by Spartan engineers were tearing down the battlements and walls of Athens itself so that her people, and her former allies, must be defenseless before Spartan military rule.

But the hegemony of Greece was not yet decided. Unholy alliances were still possible to keep the causes of lesser confederacies alive, and friends of Athens called for Persian help. The war smoldered on. But, what was needed to end it was a new technical trick to set against the conventional hoplite army of one side or the other, for it is always improved technique that brings decisive victory. Tricks were indeed coming, the first from the little state of Thebes, whose armies rose in a few short years from semiobscurity to near-dominion over Greece.

The Theban adventure seems to have been made possible by the genius of one man, Epaminondas, one of the greatest of all captains. He designed a new array of hoplite soldiers to force the end of a Spartan line, letting him roll up a phalanx from one flank, and he had the necessary flair and unflappability to vary the new design to each different battle that he fought. His trick depended upon the fact that the classic phalanx always tended to move on the slant as each man crowded behind the shield held by his neighbor to the right. Clashing armies tended to wheel round each other. Epaminondas arranged a massive column of armor on his own left flank, at least fifty ranks deep, and saw that these men pressed forward as his main line tried merely to hold. If this heavy column could penetrate the shield wall immediately to its front, the following soldiers could then split the Spartan ranks from the end. It would, indeed, be a desperate thing for the heavily armored and cumbersome men of a shield wall to be attacked from the side in this way. Under Epaminondas the Thebans did just this to Spartan armies and quickly won a series of battles that almost made Thebes master of Greece.

But the Theban military trick was purely tactical, using established weaponry and drill, though in novel ways. Advances of this kind are quickly learned by enemies and a tactical answer found. As it happened, Sparta did not need even to do this because Epaminondas was killed as he pursued a beaten enemy in his last fight and no other Theban general had the wit to use the new massed columns of infantry with his old flair. But the seed of the eventual unification of Greece by military means had been sown by Epaminondas all the same. In his camp, and watching him work, was a twenty-year-old prince, both a royal guest and hostage for the good behavior of his people. His name was Philip, soon to be king of Macedon and father of Alexander the Great.

Epaminondas had used a massive weight of standard armored men to punch through a Spartan army on a sensitive flank. Philip re-armed the whole line to the same purpose of punching through. Longer spears, a thicker front of spears as those from rear ranks reached between their comrades in front, first-line troops in every rank drilled to turn about and defend in flank, the whole a super phalanx of more armor and more spear points in action at the same time. It was an instrument unbeatable by the conventional shield walls. With it Greece was indeed brought under one government, not by one of the classic city-states like Athens or Sparta, but by the interloping northern community of Macedon.

As the several Greek city-states fought their way into union during the last century of their existence, there were copious signs of those increasing national ambitions which reflect the desire of more people for broader niches. The equipment of a Greek soldier was expensive and his training was long, yet their society produced these dangerous luxuries in such numbers that they sold their services to foreign powers, as when Xenophon and his ten thousand intervened in a Persian civil war, and afterwards cut their way home through more than a thousand miles in one of the most celebrated marches of history. Greek orators urged the acquisition of Persian lands as necessary to Greek well-being and a speech by Isocrates, forty years before Philip made his phalanx, still survives, urging Greeks to bury their differences and conquer Asia. The Greeks themselves realized that their technically advanced, aspiring, expanding populations needed both central control and the loot that their armies could bring them.

The jealousies of rival claims meant that the necessary union had to come about by military means. Philip and his Macedonians gave the Greeks union. Now there was one nation, under one strong government, owning expansive lands, conscious of their cleverness and superiority to others, with armies trained and equipped so well that they knew no neighbors could stand against them, and with frustrated freemen to provide for as well as aspiring younger sons. In that ebullient time the cry might well have echoed that of the conquering United States a hundred years ago, "Go West young man," only for the Greeks the direction was East. Philip was killed, and he bequeathed his restless state and his

incomparable army to his gifted, learned son. The
Greek people blessed his enterprise and Alexander
went out and garnered in the Persian lands.

When the Greek armies had the Persian Empire in
their grasp something curious happened. Alexander
started behaving well toward the conquered peoples.
He let it be known that he had come to give good government, not just to plunder; he left Persian officials in
office, he insisted on Persian courtesies about his court,
and he even began to dress like a Persian himself. Here
was a young man dreaming of earthly kingship, giving
good government to all peoples, bringing the happiness, just laws and the improved quality of life, which
Greeks had invented, for the enjoyment of everyone.
There is little doubt that his intentions, once he had
conquered, were sincere. Even more than for his qualities as a general, they are what have led historians to call
him "the Great." But Alexander misunderstood the tide
of history which had brought him to his power, as he
also failed to understand the pragmatic motives of the
generals, the financiers and the soldiers who had
brought him to an imperial crown. The tide of history
was the expanding needs of the expanding populations
of Greece. This tide had forged the magnificent army
in the long trial by civil war, and this tide had sent out
as officers landless sons with fortunes to build and hum-
bler folk seeking plunder to improve their lots a little.

Alexander wanted to be the lawgiver, but his soldiers
wanted to take land from Persians. In the few years that
he lived Alexander had his way, bringing, after the car-
nage of his conquests, peace and the better styles of
living—the improved niche of the Greeks. A new way
of life, the Greek culture, and even the Greek language,

took root throughout the Persian dominions and lasted for centuries, illustrating the cardinal human trait that when people recognize a superior standard of life they emulate and adopt, rapidly and for good. A new niche can spread across whole nations in a single generation. And so it was in Persia after Alexander's conquest. But the peace which Alexander imposed was a fragile thing, swept away by the wars among his generals which followed on his death.

Many a modern historian has brooded to the point of anguish over Alexander's untimely death, wondering what might have been, if he had lived the forty years more of a long life. Would his policy of reconciliation have brought in a long golden age of peace and prosperity for all? The consensus seems to be that it might have done so. But this view mistakes the underlying purpose and cause of the Greek aggression. Numbers and aspirations were growing in Greece, and the resulting demands were to be met by Persian resources. The pressure to take over a Persia held by a strong government built up by a long-lived Alexander could not have been gainsaid by the good intentions of the lawgiver. Furthermore, Persian numbers themselves were rising, and this, indeed, had been the underlying cause of the decay of the Persian governing fabric. Greek technology may have stimulated the release of resources from this vast expanse for a time, but the breeding strategy of all the inhabitants of the empire would have remained that of rearing the large families which the new resources let them afford. The rising numbers of people would have caused repressions, discontents and civil wars, and the history of a firmly established Alexandrian Empire would have been but a prolonged version of what actually happened. But his death led to the rapid breakup of the empire and the establishment of its separate parts as kingdoms under his Macedonian generals, and these kingdoms in the course of time fell, one by one, to the superior military techniques of Rome.

Civilization was progressing in Italy, no less than in Greece, but it was influenced by the long secluded shape of Italy. There was no scarcity of good agricultural land in this long peninsula, so the Italians had less need to turn to the sea for trade or conquest. Populations could grow for a long time with no more than the local adjustment of borders between tribal states. Then the states which deemed themselves the most worthy resorted to the usual armed aggression against their immediate neighbors. Techniques of land warfare were developed early and earnestly; the primitive infantry phalanx of armored spearmen probably appeared in Italy as early as it had appeared in Greece, and when Italian fought Italian it was to jab this deadly instrument against another equally deadly. And yet the military evolution of Italy went neither in the direction of the Spartan absolute expertise in conventional war, nor to the ultimate development of the phalanx into the terrible instrument of Macedon, but into something quite different, the legion.

We of the West are brought up with the word "legion" as part of our heritage. Legions were the regiments that made Rome mistress of the world. We even name our old-soldiers' associations after the legions. But the Roman legion was not just the name given to the regiments of an Italian city; it was a new, and terrible, technology of war.

The Roman legion met the armored tank of phalanx and shield wall with fire power and dash. It was a missile-throwing cloud of men. You can get the essence of the trick by thinking of the traditional picture of the Roman soldier; a man with a very large, oblong shield, and a short, straight sword. The shield was big enough that a man did not have to huddle against his neighbor to feel safe, giving him freedom of movement. The short sword was a deadly tool for individual combat, for cut and thrust in the melee of battle. But this kind of fighting was possible only when the Roman and his mates got into the lines of a shattered enemy formation; then they could cut and butcher between the shafts of the enemy spears. Before this swordwork, however, the legionary did his main fighting with missiles; the javelins and darts of which Latin writers talk so much. Legions did not attack shield walls, they tore them down from a safe distance, out of reach of the thrusting spears.

Picture yourself holding your place in a shield wall as a legion bears down upon you. You huddle in your armor, your spear thrust forward, waiting. The Roman array coming down on you looks strangely open and loose, though with desperate discipline in its movements. There are no spears, just the short "Spanish" swords being clattered on the shields, giving rhythm to the deadly precision of the legion's steps.

The Roman front looks solid, in spite of the open spacing, because the men are arranged like the squares of a checkerboard. At twenty-five paces the front line stops, there is a brief stillness then a shout of command, a swirl in the ranks and a flight of heavy javelins thuds into your line, wounding, impaling, clamping as immovable dead weights to the fronts of shields. Another swirl and another volley. With the thud and clatter of each javelin storm come the sounds of men pierced and torn open: you try to shrink small behind your shield. The

ranks of the checkerboard to your front shuffle, and a
fresh line of throwers faces you. They shuffle yet again;
still the measured shouts of command; still the flights of
javelins thudding into armor or flesh. There are hurt
men all round you, gaps in your ranks to be closed; and
you have not struck a single blow.

Soon your line will be so torn that these nimble foe-
men will run among you, stabbing with those short
swords. Close the ranks; you must keep the ranks
closed; that, and get to grips so you can use your spear.
In desperation your officer orders the charge and you
lumber forward, clanking in your heavy armor, your
arms tired with the weight of a shield dragged down by
a javelin or two. But the checkerboard in front of you
shuffles again and melts to the rear. And there, closing
the gaps through which they have let their comrades of
javelin and dart men pass, stand the men of a Roman
shield-wall like your own; in the classic order of phalanx, heavily armored, shoulder to shoulder, spears waiting for your wounded ranks, the back-up line of the
legion.

The Romans had invented what was to be one of the
most effective military techniques of all time. Not only
did the legion combine the maximum killing effect of
missiles discharged from a safe distance with the perfect
equipment for hand-to-hand fighting, but it was versa-
tile in the field, adaptable and responsive to command.
There was much room for tactical ingenuity in the putting down of the checkerboards of swordsmen and the
blocks of phalanx. There could be variety in the pace of
movement. It was possible to change the array at short
notice to accommodate some chance of battle. It allowed
for the individual ideas of a commander in fitting weapons and tactics to a particular campaign. Time was to

prove that a Roman army organized into legions was unbeatable by any military force of antiquity, unless the legion was ineptly commanded and the opposing forces were led by a very competent general. Even terrible Macedonian formations like those with which Alexander conquered eventually were to be torn down by flights of javelins hurled by nimble legionaries. It was this technical advantage in war that was to give to Rome, rather than to Greece, dominion over the Mediterranean lands.

But for a legion to go to war it had to be trained with the most industrious and resolute care. Training mattered above all. The Romans called this training *"disciplina,"* and they meant not just blind obedience to orders, but absolute skill at arms and movement. All things necessary for the common good must be done as by second nature, from the calm response to orders in the press of battle to the wearisome digging of trench and rampart round the camp every single night; come rain, or enemy, or peace. Only through superb training could a legion be always on its guard as it marched its way through hostile country. Only discipline let men with short swords go in open order against a phalanx and win.

This training started in the days of the old Republic, as the leading citizens imposed it on themselves. The free men of the better class did their own fighting, just as the ruling males of a Greek city equipped themselves as hoplites. Roman males practiced fighting in fields round their city, the *Campus Martius,* set aside for this lifelong training in war. It is curious that we name the peaceful land of a university, the campus, after the place where the Romans learned to fight. Yet it reveals the vital fact that success in war goes to the nation that can plan, invent, and devote wealth to the military art. The conquest of barbarian land, which was to make up the bulk of the Roman Empire, was to be a perfect illustration of aggressive war waged by the wealthy and the competent against the backward and the poor.

But first Rome had to meet other wealthy and civilized states; Greece, whose Macedonian-inspired armies were to collapse with surprising suddenness before the legions, and Carthage. The Greeks met defeat with civilized resignation and lived under the new rule. For Carthage it was different.

While the Roman Republic used its legions to take for itself the barbarian lands of Italy, and the Greeks built a great nation-state, another civilization had ripened in what is now Tunis, but was then a great city called Carthage. We have no records of the early feats of the Phoenician people of Tunisia comparable to the records of Greece or Rome; the utter annihilation of everything Carthaginian by the Roman armies has seen to that. But we do know that a Carthaginian power, culture and constitution, possibly as admirable as those of Greece and Rome, had grown between the Tunisian desert and the sea. Yet the habits and culture of the Carthaginian people had been shaped by a different set of ecological realities.

Carthage was in a fertile place, but there was not much of it. The people could not win more resources by aggression on neighbors, for the neighbors held only desert. Instead, the Carthaginians took the approach of the green revolution, terracing and irrigating the land they had, making their narrow strip between the desert and the sea so green with crops that it was eventually to be the marvel of Roman visitors. But when their growing civilization needed more opportunity, a broader niche for the more enterprising, there was no way of meeting the need in their narrow patch of land. Logic says that trade and then colonies were the only practicable ways to provide for Carthage an expanded way of life. We know that these solutions were, in fact, used to the extent that they became a national way of life. The Carthaginians lived by trade westward, where there seem to have been few trade rivals, leaving their mark round Africa almost as far as the equator, and round Spain to reach northern Europe. And they planted colonies in barbarian lands as the Greek cities did.

But they did not develop an advanced technology of war. Carthaginians had few civilized states with which to fight, none alongside their home city, few near the barbarian lands which they expropriated for colonies. Good weapons, good armor, good courage, and the shield-wall approach of the primitive phalanx were all that were needed to secure the colonial lands of their earlier expansion. Carthaginians were not tried by civil war early enough to force them to give pride of place to military technology as the Greeks and Romans had had to do. Carthage had fine fleets, as befitted a trading state, but an indifferent army. When the real ecological wars came, this was to be her undoing.

So we have a nation of ambitious merchants and ebullient seamen, of wealth from trade to patronize arts and material things, of high consumption, of safe, confident, well-fed families. The numbers of people in this nation must surely grow, nor will the new sons be content with less than what their parents had. The Carthaginians were going to feel the need for yet more land overseas. We have no state statistics, no census, to tell us that the Carthaginian numbers did grow, but there is evidence enough that they must have grown in what the Carthaginians did: trade, colonies, and, at the last, outright attempts at foreign conquest. It was then that their vital interests first clashed with those of Rome.

Then were to be fought three great wars; Punic Wars as the Roman authors call them. The first was a long war of attrition over the ownership of colonial lands. In the second war, Hannibal crossed the Alps with his elephants, and the third war ended with the annihilation of the Carthaginian state and people.

Roman authors, and our school books, tell the tale of these wars as a struggle to see which state should be mistress of the world; Rome, with all Italy already in its power, or the trading city of Carthage, sweeping the Mediterranean Sea with merchants and fleets. Rome did indeed go on to conquer and enslave every nation within reach of its terrible legions. But the war was not over who should be "mistress." It was a struggle for raw survival by the civilized folk of a trading state against the resources and weapons of a continental power. I suggest that the good guys lost.

The obvious expedient for Carthaginians to meet the need for more space was to expand into Sicily, the land nearest to their home, a big fertile island in the middle of the Mediterranean Sea, but Sicilian affairs were under the firm control of Hellenic settlements from Greece. Sicilian wheat was going to support Sparta; and the fortified city of Syracuse, which defied the armies of Athens, was equally impregnable to Carthaginian armies. It was this tendency to dispute the land in Sicily that had led Carthaginian fleets to be drawn into the Greek civil wars. But the Greeks fought among themselves, and Carthage found a lodgment in southern Sicily, in spite of the Greeks. Things might then have been well enough for Carthage if the Romans too had not gone looking for living space in Sicily.

The people of the Roman city had built an Italian republic incorporating not only the land but also many of the peoples of conquered Italy. They had abundant acres both of farm and forest, but they also had ambitious people, and we know that their numbers were growing and congested. Contemporary accounts give clear enough evidence of this. Delightfully revealing is a remark of the sister of an inept and unsuccessful Roman general, P. Claudius, after his death. She said that she wished he was still alive, that he might lose more men, and make the streets less crowded. This is clear evidence that the process of drift from countryside to cities that goes with rising numbers and technology was already well advanced in Italy. The Romans cannot have been desperate for land, as the Carthaginians must have been, but expanding niches for expanding numbers would certainly require new outlets. It was already the Roman tradition to provide for adventure and change by training for war, and to meet the needs of her ambitious young men by sending out armies to gather in fresh land for them. With all Italy firmly in their grasp, this policy now led them to Sicily. There the legions met troops guarding the Carthaginian lodgment, and the First Punic War began.

Carthaginians fought on land as a simple phalanx of spearmen and swordsmen supported by squadrons of excellent light cavalry. This had been good enough for the old Carthaginian colonial policy but it was no match for a legion. It is likely that the full development of legionary tactics was yet to come, and that the final shape of the legion may have actually been honed out in these first fights with Carthage. But the technical superiority of the Roman armies seems clear enough. In a grinding war of attrition, Romans repeatedly won pitched battles, only to find beaten enemies regrouping and striking back under the direction of competent generals. And the Carthaginians kept their cause alive by the use of sea power.

Sicily was an island which could not be held or taken without command of the sea, and the Carthaginians had the first navy of the day. Again and again Roman fleets were utterly smashed by the lines of galleys and rams, handled with skill and panache by Carthaginian sailors for whom seafaring was a way of life. Then Carthage would resupply its armies. But the Romans built more ships each time they lost at sea, using economic resources that seemed without end.

The struggle went on for twenty-four years, a whole generation whose lives were spent in war. This, surely, is a measure of the issues that were at stake. Then, at last, the Romans won a sea fight, destroying most of the Carthaginian fleet then in being. The Romans had done this both by learning seafaring through hard experience and by bringing their technical ingenuity to the problems of fighting at sea. They fitted each ship with boarding ramps, which rotated about a swivel. These could be swung round to either side of a Roman ship, then dropped so that they nailed themselves to the Carthaginian decks with the great steel spikes underneath them. Then legionary soldiers, with broad shield and short sword in hand, swarmed across the ramps. It was decisive, and the Romans won the Carthaginian fleet. At once, Carthage sued for peace, even at the price of yielding Sicily to the Romans.

This first Carthaginian surrender reveals the different resources on which the warring societies were based.

The Romans could lose fleet after fleet, and always build another. The Carthaginians lost one fleet and sued for peace. Moralists have told us that this reflected the greater vigor of the Roman Republic, a sort of puritan ethic which triumphed over adversity. The Carthaginians are labeled as quitters, certainly not puritans, and with a dark hint that they might not have had much ethic about them either. What it really meant was that Rome could easily build fleets out of forests growing on their broad acres, whereas the Carthaginians owned no forests.

The Carthaginian wealth was laboriously collected by trade, and the income from trade was barely enough to sustain her people—which was why she needed land in Sicily. In a sense, her fleets were part of her capital. She was waging much of the war from capital; trading nations do. But Rome ran the war from income. Even a new fleet could be provided from income, or at least from forest lands which were not being used for something else. Of people to die, neither side had a lack. But Rome fought with the material resources and income of a continent. Carthage could meet her only with the finite capital amassed by a trading state, and with the precarious income of commerce.

The Carthaginian merchants saw their wealth sliding away in the long war of attrition, cost accounted very clearly what it was doing to their capital, and quit. It was an act of weariness and appeasement that we can understand and that was even necessary. Yet it gained them nothing. The Roman appetite was the appetite of an expanding population, and it could not long be checked by feeding it blocks of land. And the Carthaginian need was the need of a similarly expanding population, and it could not be done away with by an act of surrender.

The wiser men of Carthage saw how grievous were the pressures on their standard of living, how bleak a future might await their posterity. There were more people but no more land. Doubtless you could expand trade, but was this certain when your homeland was small and your colonial ports scattered? To the north had appeared this tremendous military power of Rome, with seemingly invincible armies ever spreading her bounds, and now a power to be reckoned with on the sea and able to send out merchants to compete for trade, with the backing to their bargaining of military force. The remaining barbarian lands were mostly civilizing or brought under the protection of Rome. Where were the crowding peoples of Carthage to turn? Already there was unrest in the populace; a discontent with their lot coupled with impatience with a ruling class which had admitted defeat in war. To this was added mutiny by mercenary troops brought back from Sicily; and for whom there was neither land nor money. What should wise policy be?

Let us suppose that Carthage had, indeed, sent her soldiers to take Sicily for the reasons a professor of ecology suggests, to seek more niche-space for increasing numbers of ambitious people; then both the desperate state to which she was reduced, the steps she took and the final events, all fall into place. On this ecological analysis, the expedition to Sicily was conceived out of need. In losing the war this need was still present. But now there were the added needs caused by diminished capital, the loss of allies, loss of the fleet, and, final acute difficulty, the problem of what to do with all those mercenary soldiers.

Probably Carthage should have gone under right away; accepting the status of offshore island to be incor-

porated and ruled by Rome. She had neither the resources nor the military technology needed to stand
against a continental power possessed of unattainable,
state-of-the-art weaponry of war. But the end was delayed because there came to the rescue of Carthage two
very remarkable men, Hamilcar Barca and his great
son, Hannibal.

Hamilcar finished off the mutiny in the sternest of
soldierly ways and then offered soldier and citizen alike
what was a real solution to the Carthaginian needs. He
would find room for them in the barbarian lands of
faraway Spain. Carthage would find resources for its
people; its colonists and its trade in this western extremity of the Mediterranean, as far away from Rome as it
would be possible to get. This was a plan both ecologically and militarily sound. England did something like
it eighteen hundred years later, when the English
turned away from sending armies to France and spent
their energies peopling America.

And it was done. Hamilcar appears even to have
given good government to barbarian Spain, so conducting himself that there was a minimum of friction and so
that, when the next need came, the Carthaginian state
could call on Spaniards for loyal service in the armies.
But this was to be under Hannibal, who succeeded his
father as governor in Spain.

Yet even in Spain the Romans would not leave Carthage alone. By then Rome was sending trading fleets
round the Mediterranean, just as Carthage was. Their
traders were competing for markets. In the Roman Senate there were cries for protection of trade. Why should
Romans be put out of work by the ancient equivalents
of Datsuns and Volkswagens, the products of a beaten
enemy? Eventually the Romans found an excuse in a

local Spanish campaign of Hannibal's, sent an ultimatum which was deliberately meant to be unacceptable, and declared war on Carthage. The Roman aim was the utter destruction of her trade rival. With the legions to call on, it ought to have been quick and easy. But it was not—because of Hannibal.

Hannibal gathered an army in Spain and began an epic march of a thousand miles along the coast of the Mediterranean Sea, across Gaul to where the Alps waited him. This was a march through barbarian country, by an army that had to live off the land. And yet, that army did not wither, it grew. These fresh soldiers were to leave their homes, cross the high mountains and descend into the plains of Italy where they would meet the most terrible armies known to their day. It was a leader who could build an army in these circumstances with whom the Romans now had to reckon.

Hannibal had the legendary elephants with him, the one special technique of war available only to Carthage of the Mediterranean powers. These war elephants, partly armored against missiles, were intended to break the legionary lines as massed tanks are used to burst opposing armies in modern war. Not only might elephants disrupt Roman formations, but the moral effect might be more potent still because the dire need of any army facing a Roman legion was to break down the Roman discipline. Certainly having an elephant trampling above you would test your unflappability rather strongly. If the Roman nerve broke, then perhaps Hannibal's phalanx of African spearmen, and his brave but rowdy barbarian recruits, could cut through the disordered Roman lines. But it was not to be, for a thousand miles of marching and the high, snow-stormed passes of the Alps were too much for tropical animals. It is said that four elephants got through to Italy but, if so, there is no record of them in the fights that were to follow.

Forcing the Alpine passes must have been a desperate thing for an ancient army, long lines in the thin air trudging on, dragging food for elephants that yet die, a sense of worry at the outcome as the privations mounted. But Hannibal's brave spirit carried them through it all so that they came down from the passes to the North Italian plain. And the Roman Senate dispatched the legions to deal with them. It was to be legion against shield wall; interior lines against exterior lines; superior numbers and massive reserves against a small invading army far from home. But then Hannibal began to win battles; against all probability; against appalling odds, he began to win. His most famous battle, at Cannae, shows the sorts of things he did.

At Cannae, Hannibal posted his men with a river at their back, in a line between two small hills. It was a come-on to the Roman generals, for the Carthaginian army must seem to be in a trap. The legions formed and advanced, the javelins showered, the shield walls of Carthage gave before such a technically superior assault from superior numbers, as they had to; back toward the river; back in a deep U with the legions pressing after them. Then from over the shoulder of each of the flanking hills appeared the rest of Hannibal's men; the best armed and the best trained, shoulder to shoulder in clanking steel they crunched into the ends of the bending line of the legions. And from miles away came the galloping horses of Hannibal's African cavalry to dash into the Roman rear. The legions disappeared into knots of desperate men, forming circles, fighting on all sides, sword against thrusting spear. Few of them survived the day.

Cannae was one of the greatest triumphs of the human spirit over technical odds of which we have record. It is still something to do the heart good to think on, so great a triumph of the genius of one man over brutal technical and material superiority. Hannibal did nearly the same thing twice more, and he supported himself in Italy for sixteen years, prowling round the Roman city he was not strong enough to attack, but so fearful a person that Romans could not bring armies against him for years on end.

There was one provocative interlude in those sixteen years, the moment when Hannibal almost received the help of an army raised by his brother, Hasdrubal. Hasdrubal came with levies from Spain and Gaul, doubtless recruiting with the glittering promise of his brother's name. He got to Italy safely, but he was no Hannibal and the Romans knew their power against mere Carthaginian soldiers without the genius to lead them. The Consul Nero, an old man who knew how to war, intercepted the reinforcing army along a little river called the Metaurus, and saw to it that the laws of technical superiority were obeyed. Hannibal learned of how close his reinforcement had come when his brother's head was hurled one night into the light of the fire before his tent. But neither Nero nor his soldiers dared try their skill against Hannibal himself.

Many historians have wondered what Hannibal was doing those sixteen years in Italy. He just maintained his small army without ever striking a blow which should win him the war for good and all. What was it all for? But when we know the ecological realities that faced Carthage, it is not hard to see. Hannibal had to set the policy of his island nation when it was attacked by a far more powerful, continental state, possessed of un-

145

matched armies, and demanding that Carthage hand over all that made a happy life possible for her people. What then should be Hannibal's policy and aim? It surely is obvious. He had to make the Romans agree to let Carthage be free to trade and settle in Spain.

Hannibal's position was like, but not so simple, to that of Britain's Churchill in 1940. On his first day as Prime Minister, Churchill had said in the House of Commons, "You ask, what is our policy? I will say: It is to wage war. . . . You ask, what is our aim? I can answer in one word: Victory." In a later speech Churchill showed how this "victory" would be won when he told of fighting on until "The New World with all its power and might comes to the rescue and succour of the Old." Carthage had no friendly United States; nothing except trade, the colonies in Spain, an army of loyal barbarians not up to the latest technology, and Hannibal's brave spirit. So how would Hannibal have answered Churchill's questions? His policy? Likewise to wage war, for he had no choice. But his aim! Victory? This surely was impossible. So what could his aim be? It was the old one of "Let my people go!"

The years in Italy were spent sending the message, "Let my people go or I will hurt you." This was Hannibal's aim: to hurt and stay alive until Rome gave Carthaginians the right to live in their own land and in Spain, and to support themselves by trade. Hannibal's celebrated messages to the people of Italy clearly support this aim: "My quarrel is not with you, but I come to fight for you and against the city and armies of Rome." These messages spoke the simple truth.

But the Romans always knew their strength, and they would not let the Punic peoples of Carthage go. One man's genius could not hold out against technology and odds forever. At last the Romans found themselves a general of Hannibal's own stature, Scipio, whom they afterwards called "Africanus." Scipio drew Hannibal after him to the defense of Carthage itself, in Africa. Then, at the battle of Zama, he made no mistake with the handling of his superior legions and saw that the laws of technical advantage were obeyed, Hannibal or no Hannibal.

Hannibal did all with his inferior military equipment that a man of genius could. He had to receive a javelin-throwing cloudlike legion with a shield wall because that was all he had, so he drew his army up in three separate shield walls, each stronger than the last, each formed by men made resolute with the knowledge that their homes, their wives, their mothers, their children and their hopes, depended on killing the Roman soldiers who would come against them.

And this time Hannibal was able to try the one special weapon which his African heritage had given him, elephants. In front of the first shield wall were eighty war elephants, armored and carrying missile-throwing soldiers. Surely not even a cloudlike legion could reform after the terrible passage of such a charge as these could deliver. But Hannibal's elephants had no more fortune against Scipio's legions than had Darius's war chariots against the phalanx of Alexander. As the elephants started they were greeted with the clamorous blasts of trumpets and cymbals sounded from the Roman ranks on a given signal. Then nimble men hurled stinging missiles, so that the pain and noise in front was too much for the poor beasts. They turned and fled this monstrous usage, trampling the lines of their masters as they went.

Then the battle became scientific killing. The legions

took the shield walls one by one, cutting them open with their flying missiles, then launching their nimble ranks through the gaps to cut down the Africans piecemeal. When the last rank was reached Scipio actually drew off his troops for a while, rested them, saw that their missiles were restocked, and arranged his line to be just the right length for the proper reaping. Hannibal and his men stood as the French guard later stood at Waterloo, without fear but without hope. And then, shortly, most of them were dead, and the Carthaginian resistance to the Roman greed for land was over.

The government of Carthage surrendered unconditionally to General Scipio. He could have plundered everything they had, but he, soldierly and statesmanlike as he was, wanted only reconciliation and a just peace. "Stop building warships, never tame another elephant for war, build no army against us, and we will let you alone to trade peacefully from your city." Such were Scipio's intentions. He was following the doctrines of many a soldierly writer of later times, who took as their thesis that wars were merely extreme ways of settling disputes. Once the dispute was settled wise policy did not press the beaten foe so desperately that he was moved to fight again, but left him with the means to satisfy the people's wants and even with a little dignity. But Rome and Carthage had not fought to settle a dispute. They had fought over the lands and resources needed for their growing populations. A soldier might see a dispute settled by the war, but a farsighted ruthless politician could see only the growing needs of expanding people stretching in front of them. Carthage's need for new resources was not assuaged at Zama, and Rome's was merely fed for a time.

If the Romans were to go on raising the standard of

life of their people, as they must to avoid unrest at home, and if the numbers of Romans continued to grow, there must come a time when they needed the trade outlets which belonged to Carthage, as well as the lands in Sicily and elsewhere which they had already taken. If the best intentions of Scipio had been followed, the growing needs of Carthage's people, squeezed against the even more quickly growing needs of Rome, would still have driven her to desperation within a few generations. But, as always, politicians remote from the beastly consequences of their acts saw more clearly the long-term necessities. The acerbic Cato thundered his celebrated dictum in the Roman Senate House, *delenda est Carthago* ("Carthage must be destroyed"). The political screws were tightened and the people of Carthage were shown despair within fifty years of Zama.

The Carthaginians really seem to have tried to live in peace. They were denied an army, and even personal weapons, by the articles of their surrender. They had no Sicily and no Spain. But they were merchants and skillful ones. Although their profile had to be kept low, they managed not only to support their city by trade, but even to collect a little wealth again. This had been their undoing, for wealth invites piracy and robbers; trade caravans are natural prey. The merchants had to convoy their traders and the convoy had to be armed. So the Carthaginian merchants armed their traders. This was a legal infringement of the treaty of surrender and it was all the anti-Carthage lobby in the Senate needed.

A legion and an ambassador disembarked from the galleys on the beach near Carthage town. "Bring your weapons out of the city and pile them on the beach." The people did, because they had no choice. "Now move back from the coast ten miles and live there, all of you." It was an ecological sentence of death, for there was only the desert behind. The last free merchants of Carthage knew that it was better to die on their feet than face so lingering an existence or slavery in Rome. They bolted their city gates, weaponless as they were, and fought. It was one of the greatest defenses of all time, for it took the legions a whole year to break that city. The Carthaginian smiths forged, the women gave their long hair to fashion the springs of ballistas, they all fought. But eventually the Roman soldiers stormed in, from flaming house to flaming house, enraged by the fear of so brave a resistance.

This time the acts of the legions were driven by ruthless political will. When the young Carthaginian men had been killed, their city was taken apart brick by brick, their irrigation works were torn open, their crops were burned, salt was symbolically scattered in their pastures, and their survivors were driven out into the deserts. We do not know just what proportion of the people were actually killed by Roman swords, but we have the Roman boasts about how thoroughly they eliminated every trace of the works of their ancient rival. Probably a large portion of the people were left to get on as best they could in a wasted land. Very many would have died from the resulting privation, and few of the rest would have been able to look after their children. There would have been a generation failure, as the people who died of old age were not replaced by a younger generation. So the actual near-extinction of the Carthaginian people was a drawn-out process of several years, rather than a quick bloody slaughter, but the result was that expanding numbers of Romans could eventually enjoy all the resources of trade and land which had once made happy these other civilized folk. The ruthless logic of a war of need had been carried through to the end. Those who think that aggressive wars are just methods of settling disputes should ponder well the fate of Carthage.

Once the Italians under Roman government had taken Sicily, and all that had belonged to Carthage, they must have had all the land and resources which their people would need for several generations. Circumstantial evidence supports this. It became Roman policy, for instance, to settle their veteran soldiers with their families on farms in various parts of Italy; a tactic useful to Rome because it helped forestall possible rebellion from former enemy city states, as well as contenting such potentially dangerous citizens as veteran soldiers. But it also suggests that there was land to spare for making new farms; the population had not yet expanded to fill up the space made available by conquest of barbarian lands. Yet social habits do not easily change. The free citizens of Rome, which meant the rulers and the middle class of a slave-owning state, had created an army and invented an unbeatable military technique because they needed to go to war to support their standard of life. Now the need was, for a time, assuaged. But the instrument was made and the habits for war were made too. Foreign wars had proved an excellent way of finding opportunity for Roman rulers, soldiers and merchants; they gave promise to the younger sons. War also made the proletariat happy with glory and loot, and it brought regular supplies of slaves to support middle-class living, as well as plunder, which could be taxed. War was the great provider of niche-space to those in power at Rome; this is the explanation for the rapidity of the Roman conquests of all the Mediterranean lands which was to follow.

Only one power was left who might have withstood

the new Roman aggressions—the Greek power, under Macedonian kings, successor states to Alexander's Greece, and equipped with the formidable military instrument he used. Already the battles of King Pyrrhus, more than seventy years before Zama, had shown that the javelin-throwing legion would be the military answer to the phalanx. But now it was to be put to the final test. The legions met a phalanx of a Macedonian king at Cynocephalae in Greece just three years after Zama. The phalanx was poorly ordered in rough country, it withered under the shower of missiles, and its men were slaughtered. But the final test came at Pydna forty years later still. The last of the Macedonian kings, Perseus, met a marauding Roman army in his own dominions with the full and properly drilled array bequeathed his country by Philip and Alexander. The phalanx defied the legions in the open, but was drawn to rougher ground where the missiles, once again, broke it open letting the legionaries enter in the last fateful slaughter. Macedon, its wealth, its power and its king, were extinguished forever, and the Greek states became satellites of the swelling Roman Empire.

Carried on by the weight of tradition, which now decreed that aggressive war was a way of Roman life, the unbeatably deadly legions were then thrust in all directions as far as they could be supported by the logistic techniques and the communications of the times. It took only a matter of decades to do this until the Roman Empire, as we know it on the maps, was made. The limits were commonly geographic, being the sea and the desert, but over the continental expanses of Europe and Asia the limits were also some function of the resentment of abused peoples. In the great expanses the barbarian enemies moved and dodged and fought again, giving no opportunity to the legions for a decisive kill. There came days of reverses. The Saxon Arminius lured three legions into thick German woodland where they could not deploy and killed their soldiers every one. Reverses such as these taught the Romans their logistic limits. And there the Empire stood, its aggressions over, glutted with space and plunder, starting a history of increased breeding and increased desires to fill up the spaces and use resources won.

These immense possessions stolen for the Roman people by the legions now gave extraordinary possibilities for the Roman way of life. Much of the Empire, the whole of what is now France for instance, had been used only for barbarian living, a pleasant enough way of life but one which ensured that populations would be much lower than could be sustained from city granaries and advanced agriculture. There was room for the younger sons to found new estates of their own; there was land to be made into farms for old soldiers back from the wars. Even with a bounding birth rate, it would be some generations before the pinch of land hunger would be felt. The Roman niche could broaden and the Roman numbers could grow for some time without serious consequence or worry. This is the fundamental reason why the Roman Empire lasted so long. The Romans were more fortunate than the Greeks of Alexander's day who had used their military superiority only to conquer filled-up civilized lands, where there was little room to expand.

But Rome owned this vast territory only because her field army was unbeatable by any military force then in existence. Alexander had been able to spread the Greek way of life among the people he conquered because their experience of earlier civilization told them it was

good. But the city ways did not seem good to free barbarians bludgeoned into the Roman state. The expanding Romans of the expanding Roman niche took away
the means for barbarian living, as surely as the European farmers of North America made it impossible for
an American Indian to live a stone age life. Again and
again, people driven to despair turned on the occupiers,
even though it always meant they would be butchered
by a legion in the end. And the turmoil, in great possessions held only by force, meant that the Romans had to
leave their government to those who wielded the military power. Via civil war and temporary arrangements
the once free Roman constitution fell, as it had to, be-
neath central dictatorship.

Although many of the Roman conquests were of
barbarian lands there were large civilized countries
scooped into the net by the legions. The Carthaginians,
of course, were declared unpersons and were driven
out to fade away, though some clung on, accepted
Roman dominion, and prospered. Some three hundred
years from the destruction of their city one Carthaginian, Severus, actually became Emperor. The Greeks
merely changed the autocrats of their own imperial
state for Roman autocrats, and went on living much as
they had lived before.

But in Palestine the Roman imposition came to people long civilized, with an ancient culture, and with both
extreme wealth and many poor people in aging cities.
For Roman invaders to seek out the good life in Palestine meant squeezing the people already there. Palestine
was not a barbarian land which Romans could settle and
rule for their own benefit but more like the heartlands
of Persia which had fallen to Alexander's Greeks. Life
of the Palestinian poor already bad, was made worse

when Romans demanded taxes. Kings like Herod and Agrippa did very well with Roman support, indulging in conspicuous consumption through palaces, processions and magnificence, while the masses of a people used to calling itself "free" felt their lives compressed.

Miserable lives in Palestine were not really the Romans' fault for life was already like that before the Romans came. Grinding poverty, social rank, and the conspicuous wealth of the few were the outcome of many generations of packing in more people into a not very fertile land. But now the Romans ruled and social woes could be blamed on the alien government of conquering soldiers. People responded to their plight in different ways. For those who would endure, there germinated a creed of acceptance of material ills which could not be righted, of counting the benefits of still being human and alive, of looking for dignity and freedom in the inner spiritual strengths of our kind. "Man shall not live by bread alone." "Render unto Caesar the things that are Caesar's." "For ye have the poor always with you." This Christian creed suggested how life might be made satisfying, even in a civilization so crowded that people were oppressed and enslaved, and the message spread as the need for it grew in other congested parts of the Empire. But for others of the fierce Jewish spirit a new creed could not satisfy. In despair they turned on their conquerors as barbarians were turning on theirs in Gaul and Britain. They fought stubbornly and long against impossible odds until, at last, Jerusalem and others of their strongholds were torn down by Roman soldiers acting as self-appointed policemen.

Everywhere the power of irresistible military force prevailed until at last there came an imposed and universal peace in the lands round the Mediterranean Sea, and it happened that a succession of clever and well-meaning men, the Antonines, became emperors of Rome. Under the Antonines Rome entered what is said to have been a golden age, a time of flourishing prosperity and universal order, a time to which many historians still look back with a longing tinged with nostalgia. The lands round the Mediterranean Sea were set apart from the rest of the world by a ring of legions, who stood to their frontier posts like a dike restraining the stormy sea of barbarians outside. And, within the dike, there was a common law, a common currency, straight, paved roads, a central waterway free from pirates, and a system of banks and credits which let commerce, industry and agriculture flourish. It seemed to citizens of those times, as it seems to citizens of the modern prosperous states of the West, that the way had been found to the perpetuation, and even improvement, of the good life for all forever.

Yet, for all the peace and stability of the Antonine years, there was much about life in the Empire which was far from admirable. There was a rigid caste system, with a social pyramid which grew ever steeper despite changes at the top as more provincial people were given Roman citizenship. The gap between rich and poor was desperately wide, and growing wider. Slaves could still be treated with a ferocity almost incomprehensible to people of our day. Growing masses of the urban poor lived without work, in disgusting tenements, on welfare payments of grain and entertained by horrible murders of prisoners and beasts in the public arenas built in every city for this necessary purpose.

And the sense of wealth was tempered also. The pooling of the Empire's resources did lead to prosperity in some ways, as when the rich agricultural regions of the Nile delta and Sicily were farmed in peace to produce huge surpluses on which people of the cities could live. Transport of grain, olive oil, and wine supported ships, harbors, sailors and longshoremen. The desires of the tiny wealthy class for luxury goods gave employment to Roman workers and supported trade with the East. And a standing army of up to two hundred thousand soldiers gave much employment to those who made weapons and armor by hand. But, in spite of these sources for imperial wealth, historians are able to talk of an almost permanent state of recession in the Roman economy from the time of the Antonine emperors onwards.

Recession showed up as a chronic failure of tax revenue. Roman governors were always hard put to pay and equip their soldiers, and they had to meet ever increasing expenses to keep the mass of the people in city tenements from rebellion by giving them free food. In the last century they actually had to meet the expenses of a true police state, paying out for a network of spies and informers. The government found no way of borrowing money, partly because the banking system remained primitive, but also because there were few large capital sums available to be borrowed. This reflected the failure of Romans to develop adequate industrial techniques. They used human labor instead of machines. It is impossible to make large cash surpluses by brute labor, so always there was little money about for governors either to borrow or to tax.

The emperors resorted to the Roman equivalent of printing money. They debased the coinage, mixing cheaper metals in with gold and silver and declaring that the new coins had the same value as the old. Our modern governments push out paper and call it "wealth"; the Romans pushed out base metal and called it "wealth"; and the result was the same. Roman coiners achieved inflation as desperate as anything seen by modern societies. What cost twenty Greek drachmas in about the year 200 is recorded in Egyptian papyri to have cost one hundred and twenty thousand drachmas just a century later. Emperors tried price and wage controls, backing them up with brutal threats not open to our governors, but it did not work. They succeeded only in ruining the middle class. At the top of the social pyramid the depressed economy made government difficult. For the mass of the people in the lower castes it made the chance for betterment hopeless.

"Recession" is not really the right word to use for these failures of the Roman economy, however, because there was growth. The Romans seem never to have stopped building roads and aqueducts and cities, nor did they fail to make weapons for armies which grew continually larger. But the growth was painfully slow. It was steady progress, not recession, even if slow enough to look like stagnation to us. But the slowness had the fatal consequence of never producing a proper surplus of capital to invest.

Yet in the earlier days the emperors did have large sums at their disposal, essentially their tax on the loot of conquest, and they could conscript innumerable slaves for public works. The imperial state did grow rather rapidly at the very first and, under the Antonines, many lived well while others hoped. Then, after only a hundred years of stability, there came most desperate trouble. It was not yet the famous trouble of attack of the Empire by hordes and nations from outside; it was attacks against Rome itself by Roman armies from

Roman provinces. The direct cause of these civil wars was the desire in each provincial army to make one of its own generals emperor. The rewards of success, and thus the real motive of those who fought, were preferential pay and the chance for loot. These marauding Roman armies were doubtless as nasty as brigands to their peaceful compatriots who lived on the line of march. But, being after nothing but loot, their fury was probably less than total when faced by other legions as well armed as they. Yet their endless skirmishings to win political power for their officers added grievously to the economic woes that followed the Antonines and lasted for a whole century.

Then came strong-man rule. It was achieved by a series of military despots, men like Severus (that first emperor of African and possibly Carthaginian descent), Diocletian and Constantine, who had a genius for the imposing of order by tyrannical force. They made a police state out of the Empire.

The smell of the police state comes down to us very clearly from the century preceding Constantine when Rome held its last sway in united government as one state under one ruler. Soldiers ruled; and feeding armies without payment became a first task of those living where the legions were stationed. Taxes increased and the bureaucracy became ever more complex. The social pyramid grew even steeper, developing into a caste system rigid perhaps beyond a real understanding. You risked your life if you spoke unconventional thoughts. Spies and informers were so prevalent that it was sometimes dangerous to talk in public at all.

Then came renewed troubles and the final, fatal wars. The Empire which had been built by force and held together by force, was finally destroyed by force; real alien force brought in by armies from outside. The Empire from the time of its wealth, therefore, went through a stable century, wars of military adventurers, a triumph of despotism, and eventual subjugation by foreign foes. In outline, this history sounds very like the general predictions of the ecological hypothesis; in detail the fit seems even better.

The written record tells of very many happenings that can only mean a progressive rise in the numbers of the people. There is evidence of population growth from the very start of the Empire, when the imperial spaces had already been won by the legions but when the old republican government was still crumbling before civil war. The dictator Sulla, who preceded the first triumvirate and the reign of Julius Caesar by only a few years, was said to have been able to find land for his demobilized soldiers in the property of the 120,000 personal enemies he had massacred. From then on there are repeated statements by the ancient historians of new towns being built by the demobilized soldiers of succeeding generations. At first these towns were but centers of farming in former barbarian lands, later they were real towns; some of them still exist as major cities of the West, particularly frontier towns like Cologne, Strasbourg and Vienna. And the ancient imperial cities got bigger too, very many of them progressively pressing their outer suburbs and city limits far beyond their old bounds and for as long as the Empire lasted. All this city growth and settlement speaks most strongly of very large increases in the numbers of the people.

And yet some recent historians have claimed that the Roman people problem was one of too few rather than too many. They point to the times of military weakness in the Empire, when it was very hard to recruit enough soldiers and when generals found it necessary to hire as mercenaries their former barbarian enemies, a proceeding which became ever more common as the final collapse came closer. They note the undoubted depopulation of large areas of countryside, particularly near the disputed frontiers. And they take heed of Roman writers who lament this drift away from the country districts or the sayings of early emperors who urged their literate supporters to have more children. Through lines of arguments like these is brought a picture of weakness coming from population loss, though how that loss can have come about is not explained.

When looked at more closely these items cited as proof of a falling population are, in fact, all of them, predicted by the ecological hypothesis which expects the numbers to grow.

Depopulating the countryside is the expected consequence of feeding large city populations through massed agriculture. It is the phenomenon well known to historians of other epochs as the "drift to the towns." Local rulers, as in Elizabethan England, pass laws against it, thinking that they need the people out there tilling the fields. But the needs of the city are not met by the traditions of agriculture in the simpler society which preceded city growth. It is efficient, and therefore cheap, to import grain in large masses from those regions which produce it best, however far away they are. Rome, and other large cities of the Empire, got their grain from Sicily, Egypt, and what is now the breadbasket of Russia in the Ukraine. It might be thought that older Roman lands would go on feeding themselves as they did in barbarian times, but the barbarian way of life had been smashed and made obsolete. The people paid Roman taxes and their youths were drawn into Roman armies or Roman cities. Many of the old farmlands were turned to the raising of meat: cattle, sheep, pigs. This takes many fewer people than raising crops and is another classic symptom of that population growth which draws people into cities. There is a market for meat; the rising standards of the better-off in cities demand it.

The drift of people from the land would, of course, be most marked in those frontier regions where the threats from warring armies made it dangerous to stay. It was largely this depopulation near the frontiers that so alarmed Roman authors because it made it difficult for armies to be fed there, even as the empty land was a lure to covetous barbarians outside the legionary dike. The fortunes of war often turned on this emptiness of frontier provinces. The truth remains, however, that a loss of people at the edges of the Empire is not only consistent with a general rise in numbers but is a predictable consequence of that rise.

The urging of Roman governors and writers to their fellow citizens to have more children has been echoed in the upper crusts of many subsequent societies; Winston Churchill made a speech to that effect in the 1950s, and modern French politics revolved around the issue for decades; there have been many states where it was a proclaimed duty to have children for the fatherland; France, Germany, U.S.S.R. The ruling classes, Roman and those since, equated numbers with strength; and they also noticed that the more solid citizens tended to have few children. Roman writers preached only at the very small section of the total population who were citizens. They were merely taking alarm, quite needlessly, at that demographic transition which is inherent in the human breeding strategy as the niche expands.

Roman writers specifically tell us that the Roman upper classes had small families, even to the extent that the great family names disappeared over the generations. This is our experience too and it is predicted by the ecological hypothesis. But the number of those better-off Romans admitted to the citizenship increased all the same, apparently more rapidly than may be accounted for by the practice of admitting selected provincials and conquered peoples to the citizenship. The emperors Augustus and Claudius each conducted a census of citizens just seventy-five years apart, showing that the numbers had grown from roughly five million to roughly six million.

Yet at the very top of the social heap it is possible that a few families were small enough to be below the replacement rate. An intriguing suggestion of this lies in the fact that none of the Antonine emperors had sons to succeed them except the last. This was a very fortunate circumstance for Rome, because these men then adopted sons to be their successors, choosing boys for their quality to be emperors themselves one day. It was probably this circumstance that gave Rome its precious hundred years of stable government in Antonine time. The good years ended when the wise Marcus Aurelius most unwisely left the Empire in the custody of a real but quite unfitted son, Commodus.

The Roman upper class kept its families down to the small size which is optimum to the wealthy, even without modern methods of birth control. Their own accounts of their sexual habits give little cause to believe that their small families were contrived through abstinence, so they must have done the trick by infanticide or abortion. There do seem to be frequent references in their literature to the practice of "exposing infants." Possibly also there was a ready market for slave women who were skilled abortionists.

The lower-middle class and the poor doubtless reared the large families suited to their niches, just as is done in all contemporary societies. From these ranks came the surplus mouths, crowded into the larger cities and which were an ever-growing tax on the welfare rolls of the Empire throughout its existence. Subsequent human experience also suggests that large families would be raised in agricultural districts, even though the land was used for livestock rearing and could support few people. Many of the children of these families would move to the cities and swell that urban proletariat living off the state. But there is intriguing evidence that the agricultural surplus of people in the later Empire may have been too large for the cities to absorb. Contemporary historians tell us of roving bandit bands of country people so large (in the tens of thousands) that once-peaceful cities had to be fortified with walls against them. We find traces of the walls still.

Slavery may have had its own special effect on the rate of population growth. The grotesque brutality of slavery in the early days may have been such that slaves had no hope of raising families at all, and it seems unlikely that people who were beaten to work or driven into mines and quarries, described as deathtraps by the Romans themselves, would leave much in the way of a posterity behind them. A shortage of slaves became chronic in Rome when fresh supplies by conquest dried up. Historians say this is because the kind Romans gave the slaves their freedom. It is more likely that the supply ran out because the slaves left few children to slave after them. So it is at least possible that population growth in the free classes was somewhat offset by population decline among slaves. The evidence for growth in the record of cities, however, suggests that any depression in total numbers through slavery was not very great.

When rulers of an ambitious civilization are faced with rising numbers, they have only three options: find more land, to conquer or for trade; increase niche-space by technical innovation; or maintain large niches for the few by repressing the mass. Romans exhausted the option of engulfing more land when the Republic became an empire and then they had small chance for trade with people outside, essentially only for luxury goods with the East. What was left was technical ingenuity or repression. Romans of the Empire avoided technical solutions, preferring slave labor to machines. After the successes of the first impact of their civilization on barbarian lands, they tried nothing new, denying themselves the option of the technical fix. Failure of technique kept the Romans poor. What was even worse, the failure also denied them the variety of niches typical of middle class living.

Rome relied on cottage workshops instead of mass-production industry, even to make weapons of war. An arms factory of the later Empire was merely a place where many slaves were put to work, beating swords and making shields. Romans sawed the planks of ships by human labor and knocked them together with nails drawn by hand. They quarried stone with hammers and chisels and hauled the blocks to the tops of high buildings with no other power than muscle. They could find employment for extraordinary numbers of the desperately poor in these ways, but they could do very little to offer the more exciting dimensions of broader niches. Engineering was scarcely a profession. Educators were almost nonexistent. Bankers and salesmen were few.

There was no journalism. Management was but the overseeing of slaves. So the trick of devising niche-space for the middle classes by technical ingenuity, which modern societies do so well, was denied the Romans.

What was left to the Roman rulers was repression, and they learned to apply that solution very well indeed. Repression was forced on them early, of course, as they enslaved their world with swords, javelins and terror. The old Roman Republic bequeathed to the Empire a state already based on slavery, repression and fear. Early reliance on social force actually helped bring about the very failure of technique which required that growing populations be held down by more force still, because a slave society of the massively poor is not likely to be technically ingenious—for why make a machine when there is cheap muscle to do the work? Romans came very close to real industry, needing, for instance, only the slightest advance to produce steam engines, but no Roman made the small tricks of invention necessary. The repressive social system stopped them by requiring slaves, and the social system itself had to be made more repressive still by default of invention.

Social repression in the later Empire tells us that the Roman possessions were crowded, as indeed they were. All the niche-space winnable by conquest, trade and slave-based industry had run out. If you were middle class or better, your children could only hope to live as well as you by elbowing someone else's children. This is real crowding in the ecological sense; all niche-spaces filled and the numbers still coming. Low population densities in parts of the Empire make no difference to this conclusion.

It is true that large areas in the later Empire, particularly at the frontiers, held few people, and even larger areas supported a population which was no more than that needed to husband livestock, but this does not mean that the Empire was not crowded. These empty spaces played Kansas and Alaska to Roman New York and Boston. By Roman technology, and in Roman circumstances of warring frontiers, these regions held their full quantum of people.

Most contemporary historians will still claim that the empty frontier provinces, and abandoned farms in Italy, show that the Empire was never "crowded." Apparently they want to reserve the term "crowded" for those regions of agricultural peasantry where people and huts dot the landscape as far as the eye can see. But they fail to relate population density to resources, as an ecologist must. Animals are crowded when no more units of niche-space can be packed into the available habitat. In this sense the record suggests very strongly that the Roman Empire was indeed crowded at the last. It was the Roman fate to become crowded at a total population density which seems very low by modern standards. They were crowded at these low numbers because they bumped into technical limits; they failed to produce any advanced technological fix to yield more resources from packed cities or marginal farm land.

The Roman ruling class had to maintain their broad, cultured niches in this empire of real ecological crowding. They oppressed the mass; but they had to give it welfare and pay for spies and soldiers to keep it in its place. These expenses of government were very high and could be paid for only by tax money wrung from the small incomes of primitive industry and agriculture. The money was never enough, so they debased the currency, produced spectacular inflation and ruined the middle class.

But forceful government, wielding a repressive and caste-ridden social system, can make a civilized state persist, even if it has pressed its numbers against the limits set by niche-space. Under their later despots, Roman governments did just this, and with great success. Art and cultivated living went on thriving in the Roman police state. In the East, centered round the Greek city of Constantinople (Byzantium), the system worked for a thousand years. Only defeat in battle prevented similar stability in the West, and for this to happen, the mighty legions had to be outclassed. A hint that they might eventually be beaten had come early from the weapons of Parthian horsemen.

Even when Julius Caesar was still sharing power with his colleagues of the First Triumvirate, and when the Roman armies were completing their greatest conquests, an array of legions was all but destroyed when sent against Parthia in the East. This was the defeat of the legions under Crassus at the battle of Carrhae. It was not an affair of Romans coming against an enemy general of genius like Hannibal, though it does seem that their own general was less than inspired. The legions were defeated by an army whose organization and weapons were quite different from their own.

The Parthians at Carrhae fought as archers on horseback. The bows they used were powerful, recurved and bonded like modern bows, laminated from horn and wood instead of plastic, but not much less powerful than the modern copy for all that. The legions marched against them on foot, long treks through arid lands, their own scouting horsemen fanning out in front of the marching infantry to search and give warning of enemies. The cavalry scouts doubtless saw bodies of Parthian horse quietly retreating before the legions, monitoring the Roman advance. Doubtless also there were skirmishes between the rival cavalry far in front of the Roman foot, perhaps a challenge from a Roman patrol, a dashing forward with lances pointed, and a retreat at the gallop by the Parthian scouts. Then the Roman lancers got their first ugly taste of what these Parthian horse archers could do as one twisted in his saddle to shoot backwards at his pursuer with uncanny accuracy. Many a Roman lancer took an arrow in the belly in this way so that the backward bow shot from a fleeing horse gained dreadful fame as the "Parthian shot."

Watched by these Parthian horsemen, the legions tramped on in their methodical, experienced and purposeful way. By night they built their rampart, ditch and palisade, sleeping safely however many bowmen the local countryside might hide. But there came the day, after the legions had walked to Carrhae, when there were horsemen behind as well as in front, thousands upon thousands of them, perhaps showing their presence by the rising dust clouds from the pounding hoofs, a sight like that which Custer's Seventh Cavalry saw on the morning of the Little Big Horn. But the legionaries were numerous, battle-hardened, secure in the knowledge of their own supremacy. They must have formed for battle with few fears about the result. Yet nearly three quarters of them were to be killed or taken that day.

The Parthians fought as dashing squadrons, running at a horse's speed round the slow-moving Roman formations. And as their horses ran, so the riders shot, accurately. The javelin men and the dartmen of the front ranks of the legions were outranged by the arrows from the composite bows. The heavily armored spearmen of the legion's backup ranks doubtless never had a chance to use their weapons at all.

The legions must have stood because they could do

169

little else. Their discipline and training doubtless held their ranks together for many hours, closing shoulder to shoulder over comrades stuck with arrows, locking their oblong shields against the arrow storm, cursing and wishing that night would come. But the Parthian squadrons had prepared for that battle in a way that was seldom possible for the primitive command systems of their wild society. The ten thousand horse archers, who swept round the legionary squares in clouds of dust and arrows, had behind them a thousand camels laden with reserve supplies of arrows. The legions had been forced to stand where there was no retreat, and ammunition had been laid by to kill as many Roman soldiers as there were from the safety of swiftly moving horses.

An army of horse archers, backed by a munition train of camels; the classical legion had no answer to this in the open. But the Roman military system as a whole had answers readily enough. You did not let your legions be caught in the open by horse archers. You moved cautiously about your conquest of Parthia, entrenching, fortifying, taking towns, spreading terror and undermining the will to resist. The Parthian state was too loosely organized, with too ephemeral systems of command, and it could not field armies like that of Carrhae very often. Once the initiative passed again to Roman hands, the generals of the legions were able to see to it that they never faced armies like that of Carrhae again, at least until the later days of the Empire when other Asian governments conscripted their horsemen with more deliberation.

Yet Carrhae had shown that the technical superiority of the legion would not remain absolute. Eventually Roman armies would have to evolve to counter the threat of horse archers and whatever other technical tricks might be brought against them. But a main purpose of later Roman armies was to defend and police imperial spaces. They always had the need for infantry, for foot soldiers to man fortresses and towns, to dig defenses, to occupy and hold territory. For these purposes the classical legionary soldier was ideal. Modern warring nations still know that you must have infantry to hold the ground even though you use tanks and aircraft to win it. The Roman enlisted man or conscript, who had replaced the old well-to-do citizen of republican days in the ranks, served to hold the imperial frontier for centuries, the essential framework of the Roman peace.

To cope with horse archers, the Roman generals took to recruiting Asiatic tribesmen of the horse-archer tradition to serve as mercenaries, auxiliary to the Roman infantry. This became a very important practice in later centuries when attacks from the East, organized by stronger governments now bent on their own aggressions against the fringe provinces of the Empire, became frequent and hard to repel. With a balanced array of line-of-battle legions and mercenary squadrons of bowmen on horseback, the Empire just about held its own in a century of struggle in the East. Yet it is noteworthy that the vital bowmen were not produced by Rome herself but had to be bought from outside. Technique within the empire was no longer expanding even in warlike things, which rather suggests that real opportunity to improve one's lot through soldiering no longer existed.

Then came a military threat which proved to be even more dangerous than the horse archer because it involved new organization as well as new weaponry. There developed in what is now Iran something remarkably like the knighthood of the Middle Ages, armored horsemen who could be ranked for a massed charge.

The idea of giving armor to horse and man is an Eastern idea, and it is an old one. The armored cavalry soldier was given the name "cataphract." There were cataphracts in the Persian army that Alexander met at Arbela, though they seem to have been kept away from the main phalanx of Alexander's army in skirmishes with the Greek cavalry screen. In those early days the cataphracts seem always to have been few in number and used in small groups. Doubtless the cost and logistics of supporting horse and man in armor on long campaigns restricted their use. But Iranian monarchs in the third century of the Roman Empire solved the problems of making armor for a whole army of heavy cavalry and of keeping them in the field. Rome suffered desperately from armies of mailed horsemen bent on returning the Eastern provinces to rule by Eastern empires. Roman soldiers had to imitate this new threat or be beaten. Under Gallienus they created a mobile reserve of heavy cavalry of their own. Of this force, the modern historian Michael Grant writes "They included heavy Persian style cavalry, looking like knights of the middle ages in the conical Iranian helmets, which the Germans later inherited; and an almost medieval concept of knighthood was to be seen in the hereditary gold ring granted to the sons of its centurions."

The Roman army was still based on the legion, but supplemented by mercenary horse archers and backed by squadrons on call of heavy armored cavalry. But we do not know the proportions of these various troops, though the indications are very strong that fighting on foot in legionary array was still routine in police engage-

ments at the frontier. Roman soldiers had a slang word
for the armored horsemen, "oven boys." Perhaps there
is something in that slang word to downgrade this new-
fangled invention, an attitude similar to that of British
cavalrymen to the tank in the years between the two
world wars. Very likely the Roman economy could not
equip and support enough mailed squadrons to support
all the legions. Certain it is that in the decisive battles
that destroyed the Empire of the West, the decisive
knightly array was launched not from Rome, but against
Roman infantry by the enemies of Rome. Yet it was not
Persians who wore the armor, but Germans.

There had always been friction where there were
German tribes on the frontier. These were free barbarians, irked by the impositions of Roman armies. They
had long memories of bitterness and insult. They saw
thinly peopled frontier lands to settle, and isolated settlements to raid for loot. And the German numbers
were growing; the Latin authors tell us so.

It was natural that the Germanic tribes should learn
to avoid legions and it was equally natural that they
should learn civilized techniques of warfare from the
city people they fought. So the German attacks grew
ever more menacing. And then new German tribes
began to appear, pressed toward Rome by great population movements from the Asiatic steppe. Among those
tribes were the Goths.

The Gothic peoples were learning war both from the
legions who checked their advance and from the cavalry
of Oriental powers to the east. They were barbaric still,
organized in agricultural tribes, equipped with horses
and draft animals for their periodic moves. They took
to helping themselves to what they needed over the
frontiers between Greece and the Danube, not only pil-

173

laging but actually settling farther and farther into what was once imperial land. And they made armor, sewing metal plates to leather clothing and trading for chain mail. They were good fighting material, and many were recruited as mercenaries in the Empire's service.

A time came when an emperor, Valens, was able to put together a powerful army to teach these Goths a lesson, wishing, so it seems, not only to pacify a province but to gain personal glory as a general. Valens' army was mostly infantry in the classic tradition, but it hurried in its advance to catch the plundering Gothic horsemen before they had time to retreat across the frontier as was their usual custom. The legions found the Goths near Adrianople. Valens marched his army early, far and fast one hot summer day until they were far in the open and also thirsty. But they came up with the Goths whose ambassadors parleyed for a while, perhaps to win time for the Gothic soldiers to array themselves and their horses in armor.

The legions stood in an open plain. The masses of armored Gothic horsemen were perforce formed up to face them. Our accounts of the battle are very sketchy, but it is clear that massed armor and mobility gave the Goths the victory. That the Goths charged is certain, and we can sense a mass of iron, contemptuous of hand-thrown missiles rattling among them, lumbering down on the legions. Imagine young Roman soldiers standing, short sword in hand, in their checkerboard pattern, on that open field, thirsty, and with those thousands of armored horsemen thundering down upon them. Perhaps the disparity in weapons was not so great as this impression grants; we really do not know. But we do know that when night came the Emperor and the men of the last legions were dead on the grass.

Just thirty-two years after this battle at Adrianople, Rome itself was taken by Gothic horsemen. Especially revealing is the way Rome itself finally fell, for the city capitulated to a Gothic army after a siege of only one week. From a purely military point of view, so great a city should not fall so easily. The Gothic armored horsemen, who could trample over a legion drawn up in the open, were hardly likely to be effective against walls defended by missile throwing machines, or among houses. But what the Goths did was to blockade the river Tiber, and in doing so stopped the grain ships from Africa from getting through.

The Roman city had become completely dependent on food grown far away. All the big cities of the Empire, with their physically, as well as ecologically crowded populations of poor, were dependent on imports in a like manner. The people of these cities did not, in an ecological sense, live where their homes were at all; they lived on those far patches of countryside which supplied their wants, just as the people of modern Britain live in the far countries at the end of their trade routes, and the people of Manhattan Island live in the American Midwest. The Empire was destroyed when those attacking it had sufficient military power to interrupt the transport of necessities for life.

And so I offer a new explanation for the decline and fall of the Roman Empire. Resources that could be extracted by contemporary technique from the lands the Empire held were not sufficient to offer the broad niche of the middle-class life of a Mediterranean city-state to very many. Rising numbers held within the Empire in cities were a drain on what little economic surplus the Empire could produce, both from the direct needs of

175

welfare payments and from the costs of the police apparatus needed to control these masses in poverty. There was neither the hope nor the surplus wealth needed for a large army of high-technology soldiers, as there had been in the days of the old Greek and Roman republics when the expansion started. The defenses of the long frontiers faltered and then distant peoples, needing land rather than civilization, pressed their war bands equipped with contemporary armament into the border provinces.

Although the Empire of the West was struck down by force of arms in this way, the real defeat can still be read in terms of breeding strategy, numbers, and niche. The growth of Greek and Roman city-states into an empire was fueled by expanding niche and the promise of more. The resulting empire filled with people until the promise of expanding niche could not be met—a direct consequence of the Roman failure to develop techniques which could extract resources fast enough to provide large niches for many. In this sense, the Empire was crowded, and the fact that there were thinly settled provinces is irrelevant. Having relapsed into a police state with very high maintenance costs, the Empire found it hard to defend itself. And it fell.

With the Roman power gone, the Gothic tribes spread into every country on the European side of the Mediterranean. Some even crossed into Africa. They came with their wives and children to stay, a mass migration of people, for they were under the pressure of crowding in their traditional lands. These traditional lands, in turn, were being pressed upon by other overcrowded barbaric and nomadic peoples from the central Asian steppes. The need for land by these technocratic Gothic barbarians in the suits of armor was apparently overwhelmingly great and they found it in the territories which had once been Roman.

They found some land thinly populated because abandoned in strife, or because it was awkwardly placed to supply the needs of cities. Other lands had few people because given over to cattle raising, which needs only a small local population of animal caretakers. The really good farmlands were organized to feed distant cities and so were lived on by populations much smaller than those they actually fed. All these lands the barbarians could settle and use directly, leaving the Roman townsfolk to privation.

To imagine the effect of this arrival of the barbarian settlers in a Roman city, think of leaving the people of Manhattan Island to fend for themselves; or of a modern British island fighting a war in which it failed to break a submarine blockade. The food supply of the Roman cities must have failed, though not at once because the breakdown of the Roman system was a protracted event. City people would get out into the countryside, to fend for themselves as best as may be. There was no sudden cataclysm, but rather a continued adversity. Roman cities and support systems collapsed slowly across decades of privation, which forced the Roman population down—through a failure to breed.

The 1970s have shown us what can happen to a city population when its people have to take to the countryside; we have the example of the evacuation of Phnom Penh in Southeast Asia, which induced such suffering that large numbers died, let alone being able to raise children. The Gothic barbarians probably never had so ruthless an impact as that, but the impact was permanent. Large sections of the Roman population certainly had to fend for themselves without any access to the distant lands which used to feed them, and with the best local lands taken by the invaders. Child rearing in these circumstances cannot have had a high success rate, even when the workings of the human breeding strategy let couples start the enterprise at all. Simple failure to breed successfully would account for a very large fall in the population, and this would have brought the numbers of people down to the carrying capacity of the local, semibarbaric communities and government that the invaders brought.

But there were doubtless depopulations of intent also. Some of the land the barbarians wanted must have been well settled by peasant farmers already and the invaders wanted uncluttered space; they must unclutter it. We have no way of telling just how strong was this component of the deliberate removal of people, but there are some suggestive clues. The ferocity of barbarians with names like Goths and Vandals still leaves its mark in our literature, as it marked the Christian writings of those times, suggesting that much killing went along with their coming. We are told that when cities fell the inhabitants were put to the sword. And, perhaps more telling still, is the frequency with which we are told that this or that people were "driven out." What does "driven out" mean? It means being denied the right to live. It means a lingering existence of privation, with exposure to all the accidents of disease and malnutrition. It means an early death. It means being without the wherewithal to save your children. It means the failure of a generation to replace itself. Some nations in Stalin's Russia may have gone this way, though modern conquests are usually too short-lived for the full effect. But in Roman Britain, in Gaul and in northern Italy, over much of what once was the western Empire, the original populations were cut down to some portion of their former size within a few generations, and their land was occupied by the swelling numbers of their conquerors.

The barbarians had come to live in a barbaric way, but they found that they had conquered countries laid out for the life of settled agriculture, and with fine buildings which it was sometimes tempting to use. For generations their ancestors had been able to see what Rome had, and sometimes to plunder a little of it for themselves. Now all was theirs. They began to settle on their conquests rather than to wander around them. And they had, from the start, to fight to keep what they had won lest it be taken from them by other barbarians, like the Huns, pressing on their heels from a population crush in the steppes to the east. Their armored horsemen found themselves defending the very lines which the legions had once held. And behind the new dike of mailed horsemen they began forging the Christian and feudal kingdoms out of which were to grow the modern civilization of the West.

But in North Africa things were very different. The fertile grain fields which had once fed the city of Rome were but patches of land on the edge of vast and unproductive deserts. And in the deserts and semideserts in a great arc of land, from the edge of the old Persian Empire in Asia to the Atlantic Ocean, were the wandering barbarian tribes who had always lived there. These people had pressed hard upon their resources since long before the coastal folk in Tunis built the civilization of Carthage, and their ways of life had changed very little. They were always warlike, because the necessities for life were so scarce that they must ever be ready to defend what they had. But they had never been able to develop anything like the formidable weaponry of the barbarians north of Rome, nor, indeed, any military technology beyond that suited for the swift raid across the desert.

They traveled light on nimble horses, without armor, and fought in swift onslaughts when the blood was up, with lances and swords. They had never been able to stand against the regular soldiers of Rome, or Carthage before her, though they had been a nuisance to both states and had also sometimes been employed by both as auxiliary cavalry. They seemed in no condition to do with the Roman Empire of the south what the armored Gothic barbarians had done to the Roman Empire of the north. Yet the crowding of some of these people into wretchedness in the country of Arabia was preparing the way for fresh aggressive wars of conquest with results no less remarkable.

When Muhammad lived, infanticide was among the normal necessities of living for his people. We know this, because particular emphasis was given in his teachings to the abolition of the practice. Infanticide is, of course, a device for keeping the number of children down to what a couple can afford to rear. If infanticide is so common as to be remarked, the suggestion is very strong that the people are pressing closely on their resources, that life is a subsistence struggle, that hope for the ambitious youth must be very slight. This appears to have been the condition of Arabia in Muhammad's time. Life would have been ordered, with people preserving most of the ancient habits which let them live in their desert world, particularly keeping probably those courteous Arab manners, so necessary an adaptation to a life that is harsh. But nearby were more civilized, more productive lands, formerly ruled from Rome but now breaking into little states or being reconquered by armies from Constantinople. Living alongside the remains of empire, crowded but independent and with their culture intact; such seems to have been the state of Arabia when Muhammad lived. Then came the prophet with his message of a new way of doing things: a new niche.

Muhammad offered the doing away with infanticide and many other boons, both spiritual and temporal, to those who followed his teachings. He also offered disciplined self-denial, which is so frequently a mark of pride in associations of people. And he offered action and fighting to the young men of an active society now in straitened circumstances. It was a mixture such as has often drawn the allegiance of the young and bold, though most captains who offer it lead their young followers only to privation and disillusion. Muhammad was different. An ecologist would say that his recipe worked because the fighting won plunder and living space, which would grant the earthly reward even as the new religion looked after the spirit.

The eruption of warriors from the desert, which Muhammad uncorked like a genie from his Arabian bottle, was yet another of the wars of aggression started because the people needed land. There seems to have been no new military technology behind this aggression; the faithful merely fought as clouds of gallant horsemen, as they had always done. But then there was no longer a phalanx or a legion in North Africa to withstand them. As the Arabians began to succeed, the tribes of all the deserts joined them, with what perhaps may be best described as holy glee. The Christian populous societies of the fertile patches along the coast had not the spirit, or the organized purpose, to stand against this deliberate fury. They were conquered, oppressed, enslaved, and sometimes subjected to that final solution of "being driven out." Their onetime resources went to support the swelling numbers and swelling desires of the people who had embraced this new niche called "Islam." In about a century all the former African possessions of Rome were in the power, and under the command, of the once barbarians of the desert, who were taking over the settlements in their own way and building from them an entirely new civilization.

Victorious military instruments do not go out of use when once their original purpose is attained. Just as the Roman legions had rolled on and on after the conquests that were really necessary to republican Rome, so the Moslem horsemen rolled on and on. They crossed into Europe at both ends of the Mediterranean; to win the world for their faith, no doubt, but also because every victory gave them more plunder, more land, more resources for a growing people. The desires and ambitions of the people of this new civilization were fed by the fruits of conquering progress.

But the Moslem military technique remained that of the desert war; the furious screaming charge of utterly fearless horsemen. This technique could not be expected to prevail against careful preparations, armor and fortifications; and so it turned out. In the East the waves of Moslem fury broke and were shattered against the fortress of Constantinople. The city was massively fortified and provisioned, and it was defended by a levelheaded soldier-king: Leo. Leo made all the right moves; with sallies, both by sea and land, when it was safe, and quick retreats within the walls when it was not. The besieging Moslem army had no way of breaking in, and it was the besiegers who starved in a long cold winter and not the besieged. Seven hundred years were to pass before a Moslem army could take Constantinople, by which time they had cannon to open a passage through the walls. Until then Constantinople was to prevent any advance of Islam into Eastern Europe.

In the West the Moslem horsemen plundered their way across Spain, taking their wives and children with them, and settling them as they went; evidence enough that there were people to spare in the swelling populations of African Islam. Then they went on their way into France, taking cities, killing all the inhabitants, loading their valuable possessions onto mules and carts to carry with them, but having too their wives and children close behind that they might settle when the fortune of war gave them permanent ownership of a nice uncluttered land.

But in France there was a Frankish king, cool enough to collect his forces and meet a charging mass of unarmored horsemen as they should be met. Charles the Hammer put together a feudal force of heavily armored men and stood them before the city of Tours. They seem to have been mostly foot soldiers, but given heavy protective body armor and surely supported by squadrons of feudal armored knights. Horsemen without armor, and not organized for massed arrow bombardment like the Parthians of Carrae, could only impale themselves on the spears of such an array. And so it proved. When Charles hammered Islam at Tours, the Moslem conquest in the West was stopped for good.

After Leo had defended Constantinople and Charles had defended France, the Mediterranean Sea became a division between the Moslem peoples of North Africa and the Christian peoples of Europe. Both civilizations were to develop in their own way, and although they were often to fight one another, the essential boundary between their lands was never changed. The peoples round that land-locked sea would never again be linked by common laws and common languages. The Mediterranean Episode was over.

CHAPTER SIX

HUMAN LEMMINGS: THE ARMY THAT GENGHIS LED

FROM TIME to time the settled nations of Asia and Europe have been beset by hordes of armed men riding at them from remote regions to the northeast. The hordes have been led by soldiers whose names echo for generations from the terror they inspired: names like Attila, Bayan and Genghis Khan. It was pressure from one of these armed movements that set the German tribes onto the Roman Empire in that last fatal thrust of theirs.

The hordes came as ferocious waves of horse-archers and horse-spearmen, swiftly moving instruments of war which killed pitilessly to gain their way and seemed to the poor victim to be driven only by blood lust and greed for what they could steal. They came out of the dry grasslands at the center of the continent, the steppes. They came with flocks and herds, which served them as a moving commissariat, and they often brought their families in their train, all borne on a multitude of horses. They have always come without warning. Their attacks are, perhaps, the most blatant and unprovoked aggressions of which we have record. They must be explained by any hypothesis which claims to give a general explanation of the causes of war.

We know most about the attacks which were most recent, about those of the Mongols and their Genghis Khan and his successors between six and seven hundred years ago, and about the Huns under Attila who cut into the crumbling remnants of the Roman Empire and into the young Gothic kingdoms which were growing out of the once Roman lands. But we have evidence for the intermittent irruption of nomad armies from the steppes for as far back as our written records show. One of the earliest series of battles to be found in our histories of warfare came about when a horde called Hyksos, "The Princes of the Desert," came rattling in horse-drawn chariots from the direction of the Caucasian steppe to cross Asia Minor and tumble on the Egypt of the Pharaohs. This was 3,600 years ago, and there are hints in our reconstruction of even earlier history of other unexpected assaults by people who came out of the desert or the steppes.

In the thousand years that followed the assault of the chariot-borne Princes of the Desert on Egypt there is evidence for at least one major irruption into Greece, but our records of those early days are very scanty. Then, 2,600 years ago, the Scythians made their practical debut into history with annihilating conquests in

Eastern Europe, and from then on the written records, at least of the settled folk who were attacked, are tolerably complete. The afflictions seemed to come in waves, there being a few generations when the soldiers of civilized states had to fight again and again for the existence of their countries against armies of people who appeared from regions beyond the knowledge of their geography. A century or two of repose would follow, when the wounded civilizations could rebuild their shattered confidence and shattered frontiers, or win back the border territories which the horsemen had entirely taken from them. Attacks and counterattacks came in cycles, a flow and an ebb of savage armies out of the sea of steppes. To Europeans the main floods of the last three thousand years have been associated with the names of Scythians, Sarmatians, Huns, Magyars and Mongols, though there were many other tribes and races involved in the mass movements. A simplified scheme of these irruptions is given in the following Table.

DATES OF THE PRINCIPAL NOMAD IRRUPTIONS INTO EASTERN EUROPE

Years ago	Name of Principal hordes
2,600	Scythians
2,100	Sarmatians
1,500	Huns
1,000	Magyars
600	Mongols

The sense of periodicity which a simplified table such as this gives is even more striking when the history of assaults on China and India is compared with it. They too were struck by Mongols, Huns and the rest, and at the same time. The great lozenge of land which is the Eurasian steppe stretches all the way from China to Europe, having a frontier with each great center of civilization in turn. It is a single geographic feature; as it were, a giant piece of real estate jointly owned by an assembly of nomadic tribes. Conquest in Europe is only one of the possibilities open to armies reared in the steppe for they could just as soon strike southwards, or eastwards into China, and the history books show that they have frequently done so. But they have struck in these other directions, and at Europe, at roughly the same time. The flood of horsed armies out of the central steppe has flowed in all directions at once during the generations of aggression. And it has drawn back from all frontiers at once during the centuries of ebb. There seems, indeed, to be a pulse of life in the steppes which pushes out armies with a rhythm measured in centuries.

Fifty years ago, when environmental studies were in their infancy, it seemed to a number of historians and anthropologists that some simple, natural rhythm might lie behind the rhythm of the nomad armies, and they sought their answer in cycles of climates. Perhaps there were wet times and dry times on the steppes, succeeding each other in some celestial rhythm with the required frequency. Rains on the steppes would make the grass grow, the flocks and herds would multiply and the numbers of the people who lived on them would multiply also. Then comes the drought, preferably for the hypothesis at a sudden disastrous stroke, and the people see their livelihood wither away. So they pick up the spears and bows, which they always have handy, to horse and away, down upon the unsuspecting civilizations in the well-watered settled lands that border the steppe. Then the rains come again; the steppes are habitable once more, the nomads can go home, their resistance to the civilized armies round about weakens, and they are pushed back whence they came, to breed in the quietness of ensuing centuries another horde.

This hypothesis sounds naive to modern ecologists, who have long ago given up trying to explain population rhythms in animals as functions of simple climatic cycles, and it is now totally discredited. Quite apart from the intrinsically unsatisfactory sound of the explanation, the proposed climatic cycles do not exist. We now have very many records of the climate of the last several thousand years, records which have been preserved for us in the sediments of lakes and bogs. Hundreds of such sequences let us reconstruct histories of climate which have ample resolution to test the hypothesis that there have been cycles of aridity with the required frequency of five or six hundred years. There have been no rhythmic cycles of wet and dry.

Yet the cyclic ebb and flow of armies coming out of the steppes is apparently real, requiring an explanation. Historians have lately looked to social pressures themselves to provide the answer, sensing a dynamic within the nomad society which generates its own cycles. Nomad society at the time of a great conquest is held under the sway of strong rulers; soldiers, fearfully successful in war and giving strong and stable government at home because they discipline their people, exacting complete obedience. Such rule of military despots must be tyrannical, and it will be less easily endured as the glut of conquest subsides. Other rulers will arise, perhaps less able than the first, or perhaps so attracted by the comforts enjoyed by the civilized rulers they had overthrown that they lose their martial grip. The vigor of the nomad empire should thus wither, letting the steady power of civilized states reassert itself to reclaim their border lands. Nomad chieftains fall to quarreling among themselves, and the nomad empire disintegrates once more into its disparate tribes. Then, after generations of feuding, there must come a time when some chieftain more soldierly than the rest succeeds in uniting the tribes once more. If some neighboring civilization happens to be going through a time of weakness, the new general can consolidate his people with a first conquest. Suddenly the fruits of conquest seem good; the tribes all give the new general their fierce allegiance, and a new flood of armies irrupts from the steppes in quest of plunder. Thus the hypothesis of social rhythms.

As a *description* of the cycle of events which lead from one irruption to another, this account is good enough. The triumphant days of the great soldier *are* always followed by a decline as his successors lose both their ambitions and their grip. The civilized armies *do* then push the nomads back and there *is* then a long interval of disintegration and tribal feuding before a new unity is forged and a new great captain thrown up. But there is nothing in this to say that the social history so described is self-perpetuating.

Perhaps people can be glutted with conquest, retire from fighting to a peaceful old age, and let their children live in comfort on the spoils of battle so that they do not themselves have to fight. But why, then, should the martial collapse last for centuries? Even more to the point, why should approximately the same number of centuries divide the successive episodes of conquest? The scientific mind finds it hard to believe in cycles generated in this way. Intermittent bickering at the frontier, as some generations are more warlike than

others, is much more likely than savage war every few
centuries, if there is nothing more than social pressures
as a cause.

Yet the cycles do make sense if the human breeding
habits are behind them. Circumstantial evidence suggests that the steppes are crowded before each great
irruption, because the hordes who ride down on the
unsuspecting settled nations alongside come in waves.
Different hordes sometimes come, one after the other,
with but a year or two between them, suggesting that
the steppes are emptying out their people by pushing
from far inside its crowded center, so that the tribes at
the edges are ejected one by one. Only when the greatest of conquerors have effectively welded all the steppe
peoples into one fighting empire, as did Genghis Khan,
is this phenomenon of separate waves masked. Separate
hordes coming in waves do not fit tidily into the hypothesis which rests on a social dynamic, but they are just
what would be expected if what we were really witness-
ing was armed emigration from an overcrowded habitat.

It is, therefore, profitable to look into the hypothesis
that each irruption of nomads from the steppes is
caused by crowding. There are too many people for
them to be able to do all the things that nomads want to
do, and they resort to the usual expedient of equipping
themselves to take more room from others by force. But
then we must explain the cycles. The hypothesis now
requires that every few hundred years the steppes become crowded, but that between these episodes of
crowding there may be centuries when the numbers of
people remain at tolerable levels. When we do not allow
simple climatic cycles of the right frequency to control
the people's food and set the people's numbers, this may

seem a hard requirement to meet. But, in fact, the problem is a familiar one to modern ecology, and such periodic gluts of population with consequent emigrations are readily understood. We see the same sort of thing in other animals of the steppes, the desert and the north.

The best-known of rhythmic irruptions of animal populations are those of lemmings. They have given rise to stories of lemming armies on the march, of Norwegian farm lands overrun, of mass suicides as lemming columns scramble onward into the sea. The tales of mass suicide are fiction, being embroidered in the minds of people staring with wonder at a countryside alive with lemmings, but the irruptions are real enough; and they serve as a useful model for the irruptions of nomad armies from the steppe.

Lemmings are small, mouselike animals that live in the tundra, the grassy, heathy land beyond the tree line, what may be called an "arctic steppe." They are usually to be found in rather low numbers, so that you may walk all day in the tundra without seeing a sign of one. Yet, every four years or so, parts of the tundra seem to come alive with lemmings. If you go for a walk then, you will find the tundra intersected by tiny beaten paths and be conscious of small brown shapes scuttling away as you walk. The lemmings are a hundred, or even a thousand, times as common as they are in other years. But next year they will probably be gone again.

The niche of the lemming includes an element of wandering, particularly in the spring when the snow melts and the ice goes out of the arctic ponds. It is easy to see why this habit of wandering is adaptive to life in a cold and northern place with a very long winter for life must be uncertain and local extinction common.

The genetic strain that spreads survivors around is most likely to be preserved in the long run; an evolutionary version of putting eggs in many baskets. And so some of the lemmings always wander in the spring. If, then, there comes a spring when the lemming populations are a thousand times more numerous than usual, there are going to be a thousand times more wanderers. And if the lemmings live high on a mountain, as in Norway, there is but a single way to go, down the valleys toward the sea, when the wandering will take on the form of a mass-migration. Scandinavian farmlands are struck by lemming waves every few years as the wanderers come down from the hills. Likewise the flatlands round the Arctic Ocean do come alive with wandering lemmings every four years or so, when the lemming numbers are high.

So lemming "migrations" are inevitable whenever the lemming population of a large piece of real estate becomes extremely dense. But why should the numbers rise so, every few years? And why should the dense populations go away so quickly? And why should the lemming populations of such large areas be synchronized? It is when we apply the concepts of breeding strategy and niche to this problem that we can see how the human waves of emigrants can be generated in the steppe.

Lemmings can have two litters a year, one in spring and the other in midwinter. The spring litter is large, perhaps a dozen babies, but the mother may not be able to raise them in the few short weeks of an arctic summer if the weather is against her, and her babies may be caught by the first frost before they are fully grown. If this happens, the whole litter dies, but the mother has one last chance at breeding before she herself dies of old age: she may try for her second litter in the winter, in burrows and tunnels deep under the snow where it is almost warm, feeding on the frozen grass and roots of last summer's growth. This litter appears to be small, just one or two babies, but it is a nice insurance policy against failure of the effort in summer.

If the lemmings are successful with their breeding a sudden, massive multiplication of the lemming numbers must inevitably follow: multiply the base number by ten in summer, triple it next winter, multiply by ten again next summer, and there is a plague of lemmings. But then why should all the lemmings of a big piece of real estate like Norway or a chunk of Canada all be lucky with their breeding at the same time?

This problem of synchrony in lemming numbers was first thought about by ecologists fifty-or-so years ago, at the time when the nomad irruptions were thought to reflect cycles in hostile climate. People were forced out of the steppe by bad weather; lemmings were wiped out by bad weather. All that was needed was a cycle of bad weather to match the lemming cycles, knocking the numbers down and letting them breed up again. But we now know that the required cycles of weather have never existed.

We are quite certain that weather does not kill or breed lemmings in a simple way to cause the cycles. But there is a way in which the weather can be a regulator, all the same. This happens because accidents of weather have to be used by the lemming sex systems as cues to switch on their breeding activities at the right time of year, and the lemmings of large areas are taking their cue from the same patch of weather. So, whatever fortune hits one family of lemmings strikes the others equally. The weather synchronizes the lemmings, even if it does not control them.

The arctic summer is very short, often having only six weeks of the frost free weather needed to rear a large clutch of babies to the stage where they can look after themselves. But this six weeks can be used to the full only if the babies are born on day one; and that means that sexual arousal, mating, and gestation, all must be done before day one of summer. The lemmings have to be turned on in the late winter, when the world is still cold and they are still under the snow. The cue that switches on their hormone systems seems to be a snap of relatively warm days in late winter.

The change of the seasons in the arctic seems always to allow an early warm snap or so before the real summer comes, and so the lemmings always get their cue to sex. But the timing of this cue of warmth is arbitrary and uncertain in the ways that we have come to expect about the weather. Some years the lemmings are switched on at the perfect time for their babies to be born on day one of summer; in many other years they are not switched on so nicely. And the breeding success of next summer is critically dependent on the timing of this cue to sex in the late winter.

This need of the lemmings to read the weather for a cue to start courting means that all the lemmings of large areas will mate and breed in synchrony. They succeed or fail together, and the fact that lemming plagues happen over large areas is explained. Getting from there to cycles is easy.

For there to be a rhythm to the lemming plagues, the populations must be cut down as often as they are built. There is no shortage of agents of death to do the trick —flesh-eating birds, foxes, weasels, disease, malnutrition or outright starvation as the lemming hordes destroy their own range, even social shock disease in the lemming crowds. But, strangely enough, none of these scourges of the lemmings is actually necessary, present though they all are. It is enough for the lemmings to die of old age.

The life span of a lemming is only a year or eighteen months. This means that there need be only one summer of bungled breeding for the lemming population almost to vanish. Suppose the tundra is crawling with lemmings one fall, and suppose that there was enough food left under the snow to get them through the winter, but let the first warm snap of the waning winter come too late and huge numbers of babies will be born without hope of being raised before the next winter comes. The owls and the foxes will no doubt feed well again. More importantly, the parents will enter the next winter, almost without issue, with their breeding lives all but spent. The lemming horde vanishes as their old bodies run down and die.

High numbers and low are thus inherent in the niche and breeding strategy of the lemmings, and their reliance on cues coming from the weather to program their sex lives is the reason that they fluctuate in synchrony over large areas. But we still have to solve the matter of the rhythm itself. Lemming highs happen every four years or so, even though the weather itself has no cycles and cannot be the pacemaker. Weather may synchronize the cycles, but it cannot set the interval between them. The interval comes from the habits of the animal itself.

It probably takes two good years of breeding to produce a lemming high, for then the lemmings multiply a hundredfold. The years need not be side by side, although, if separated, the intervening year must be a mark-time year of moderate breeding success. But it takes only one year of breeding failure to demolish the plague and make the lemmings rare again. Two years to boom; one year to bust; those are the facts that determine the size of the cycle. On top of this is grafted the random chances that the weather will bring the mating cue too early or too late. And the result is that lemming highs appear on the average every four years. But this is *only* an average. We have been able to reconstruct the history of lemming cycles for the last 150 years from the notebooks of trappers in the fur trade, mainly from the files of the Hudson's Bay Company, and lemming highs can be anything from two to seven years apart. The average on the run of 150 years is three point something years, which rounds out comfortably to four.

The lemmings that march in Norway and Canada to the journalists' delight, therefore, are not sprung by rhythms in climate, although the vagaries of weather keep them in synchrony. It is their breeding strategy, and their niche for living in a difficult arctic place, that bring their booms and busts. And the time between boom and bust is a function of how long they live and how long they take to raise babies.

Nomadic people of the Asiatic steppe are not lemmings, in spite of my title to this chapter. But the cycles of nomad aggression can yet be generated by human niche and breeding strategy fitted to life on the semi-arctic steppe, very much as the lemming cycles are generated.

Nomads of the Asian steppes lived under conditions and in ways which give a distant echo of the lemmings. They were adapted to life where there were no trees and where there was a long hard winter. They were dependent on their animals for food, transport, clothing, and even for the covering of their houses. They

197

managed quick crops of grain when the rains came. They learned to wander far with the changing seasons, to distant pastures, to water holes. In their wanderings they were able to live off their animals, which were a moving commissariat and larder, and on grain, carried in sacks and saddle bags woven or knotted in elegant fabrics of wool of the traditions we now know from Turkoman carpets.

But the fortunes of such people were critically dependent on the weather, particularly on the amount and timing of the seasonal rains. There are wet years and dry years on the steppes, as the weather changes with that characteristic unpredictability which we all know so well. In some years the annual wandering to the alternative pastures used by the tribe were more critical than in other years. But there were no dramatic changes, either from year to year or from century to century, as the climatic hypothesis for nomad cycles would have had us believe. The weather was always steppe weather, letting the characteristic life of the steppe persist.

Because the rains were fickle, however, the niche that the people of the steppes had learned must be adapted to cope with bad years and good. In lean years, the people might travel farther than usual, relying on their ancient and accumulated knowledge of steppe geography to seek out the surviving pastures. But this would mean dispute with other tribes who also made long journeys, or who used these refuges every year. The weather of the steppe would be good or bad that season over huge areas at once, and all the tribes would be affected equally; they would all wander less in a good year, more in a bad. When the rains were poor it would often be necessary to fight for the remaining pasturage. A certain handiness with weapons must always have been an integral part of the nomad niche, even more so than for barbarians living in more certain habitats.

The breeding strategy of these nomad people would reflect, in detail, this niche of nomad living. Each couple would raise the optimum number of children that they could carry with them and nurture through the lean season. This optimum would certainly be a large number. The nomad niche held none of the restraints that make the wealthy of civilized states opt for small families, nor did nomads feel the privations of poor agricultural peasants or people in city slums who cannot afford to raise many children. With a reserve of foods in flocks and herds, and with a life style that could use children as useful labor as shepherds and beast-minders, nomad families faced few constraints. The nomad life was thus a good life for rearing children and we can expect many children in each nomad tent. Numbers were under a strong and permanent pressure to rise.

Yet the nomad niche on the steppes was one that could not support dense populations. They ate meat and a little corn, both depending on the poor productivity of the steppes for forage and grain. Niche-spaces must usually have been at a premium on the steppes. So we have a chronic shortage of niche-spaces and an abundant supply of young adults to compete for them; few jobs and many applicants. The nomad cultures must have ways of allotting niche-spaces to some and denying them to others.

In all species that feed to maturity more individuals than there can be room for as adults, competition must cull the young when they are recruited to the breeding population. No doubt the nomad habit of fighting helped, as did ritual and taboo. In these ways the nomads could be compared to our hunter-gatherer ancestors of long ago. Their niche included a cultural gate or filter through which each individual had to pass to take up one of the spaces for a breeding adult in the tribe. But nomad societies that thinned out young adults must have felt friction and strife.

All that is needed to produce a surplus of nomads across all the steppes at the same time is for some run of seasons to encourage societies everywhere to let in a few more adults, because these extra couples would each rear a new family. Perhaps a chance run of good years would do this simply by letting the people wander less so that friction between distant tribes was less, or it could be simply that the herds grew, the mares gave more milk, and the grain sacks were filled. But any small change in habit that let the average recruitment to the tribe at puberty be slightly more generous, would ensure that the steppe would be crowded in the years ahead—especially if the new laxer habits were not easily abandoned.

In this way the whole steppe would start filling with too many people, synchronously, because habit was triggered by weather just as the tundra of large areas can fill with lemmings. People grow and reproduce more slowly than lemmings, and the chances of weather that affect them take longer to work themselves out. That a nomad high should happen only every five hundred years or so by these means seems reasonable.

Population highs of nomads must now be translated into armies of aggression, which is easy. Nomadic people fight over pastures and water holes anyway; more nomads on the move means more fighting in bad years; more fighting means better attention to weapons and generals; and this means the chance of raising a real army.

A likely and, I think, predictable course of events is as follows. The rains are kind and the pastures are good all over the vast expanse of the steppe. The people thrive, there is little fighting because there is plenty of pasture to go round. Taboo and blood feud relax, so that more young adults join the breeding population and more families are raised. Luck being what it is, the rains will not continue kind, but already marginally too many people live on the steppes. They try to carry on their traditional ways all the same, living their free nomadic life, wandering where they will in quest of the good pastures. But this will mean more fighting than the average generation of nomad sees because there are too many families wandering to the traditional pastures. Skill in war will become more important to the people and a warlike chieftain is almost essential to the success of the tribe. A social momentum toward a warrior state results, but the martial air of the people does not come from some mysterious moral dynamic as social historians suggest but from the needs of too many people.

Very likely the adaptations which allow more fighting are sufficient to let the people pursue their nomadism on the steppes for another generation or two. It could very well lead to a more perfect sharing out of the pastures, so that rather larger populations could be maintained for a while. But if it should happen that the rains were kind again the consequences could be more serious. Once more the people, vividly aware of their need for soldiers, relax custom and let more youths marry. Two generations of letting more nomads breed could probably be as devastating to human life on the steppes as two generations of good lemming breeding are for rodent life on the tundras. There would be too many people.

201

As always the problem would be, not that there were more people than could be fed, but that there would be more people than could be provided with a traditional way of life. Other peoples try to solve this dilemma by technical fixes, trade, social repression and conquest. Of these options, only conquest was open to the nomads. Furthermore, in the years of increasing crowding, the people have grown accustomed to ever more fighting in defense of the traditional habits. The army is already abuilding, and so is the will to use it. A period of internal fighting throws up, at last, a great captain who will offer to solve everyone's problem with a foreign war, and one of the great nomad aggressions is born.

Population cycles of steppe-people as well as of lemmings are thus synchronized by weather, although they are definitely not caused by cycles in climate. Both lemmings and people use weather as cues for behavior, the lemmings for simple sex, the people more subtly. All people behave to suit the weather but pastoral nomads are more closely tied to weather than the rest of us so that small changes in habit bring large consequences in population. Yet the climatic pattern of good years and bad is purely random, for both people and lemmings. The length of time between one population high and the next is, for both species, set by how fast each can breed, how long each lives, and how prompt each is to respond to changes in the weather. Lemmings can raise a baby in six weeks, live a year, and produce huge populations at roughly four-year intervals. People take twenty years to raise a baby, live sixty years, and produce largish populations roughly every five hundred years. The cycles are thus properties of the animals, not of climate.

When there are too many lemmings on the tundra,

the surplus must die, or fail to breed, so that the excess crop is removed. The same is true for nomads. Surplus nomads are spent as they follow their great captain in his armies to pitch their tents in border lands once held by civilized states. The steppes are relieved of their surplus people and nomadism there may revert to its traditional ways.

Fighting is no longer so necessary to the stay-at-homes and the martial needs of the people which made them submit to the triumphant discipline of their generals can be relaxed. This explains the ebb tide of nomad conquests. In the conquered territories the soldiers' families fade away, as their nomad traditions do not equip them to breed well in the circumstances of a frontier state; or else they become settlers themselves. Either way they are lost to the steppes and the armed emigration has served its population purpose. In this way does the ecological hypothesis provide a rational explanation for the periodic wars of conquest undertaken by nomadic peoples. The actual course of their conquests can be understood by following the exploits of the best known of their captains, Genghis Khan.

When Genghis Khan was a boy, the tribes on his part of the steppe fought fiercely and treacherously with one another. Fighting and war must certainly be a normal part of the nomad niche, but there is that in the accounts of Genghis Khan's early days, which tell of a ferocity unlikely to be the normal state of affairs. The boy's father was a minor chieftain who seems to have held his power by being a little more savage than those about him. He helped a local chief, only a little less petty, back to his authority when rebellion and aggression threw him out. Later, when the future conqueror was only nine years old, his father broke a journey to dine with some friendly Tartars at which benevolent dinner he was poisoned and died. His family, with the nine-year-old boy now the head of it, was left in the nomad equivalent of destitution to fend for itself by scrabbling for roots and small game as best it could. Later, in a family quarrel, the boy shot his half brother in the back with an arrow. When he was big enough, and perhaps a hard man to cross, he was able to claim the loyalty of one of his father's friends, both for a long-promised bride and for the brute material help which would set him up in the petty chieftainship he could claim by inheritance.

There have come down to us, through the writings of Persian scholars, sufficient accounts of those turbulent days on the steppes to show that the savage beginnings to the life of Genghis Khan were typical of his times. The story seems to tremble with a sense of frustration of a society made savage because the people did not know what to do. It was not that they could find no livelihood at all, but rather that their freedom of action was restricted in some mysterious way. Their culture traditionally allowed more violence than is possible for settled folk, and their frustrations stretched the traditional violence into savagery. If there were too many people for each to follow the ample ways of nomad life as it had been lived in the past, this would be expected.

The young men of the tribes were well equipped to fight their grassland wars of maneuver, surprise and treachery. Each man normally owned several horses, was accustomed to sleeping in the saddle, and could ride a week or more on a few pounds of cold dried food, so that the only stops on a deadly march were when he changed from one horse to another. They were all experts with a composite bow, a weapon to which they had been introduced when they were three years old. These bows were more powerful than the longbows of the English yeomen who cut down the charging French knights at Agincourt and Crécy a hundred years later. In addition to being superb bowmen the Mongols were skilled lancers, able to take their place in a formation of heavy cavalry. They even could clothe themselves in armor, of sorts, made from iron scales sewn to leather jackets, for the Chinese had been unwise enough to allow iron to be traded with the nomads throughout the turbulent years which preceded the rise of Genghis Khan.

When every young man was a soldier, trained by every experience of his life to the skills and tactics of steppe warfare, decisive battles could be fought only when generalship of a high order appeared. When tribes were led by the average, or the merely competent, warfare between these skillful nomad soldiers could lead only to a succession of strokes and counterstrokes, of ambush and revenge. Genghis Khan was the general who brought the decisions. He seems to have had an inspired ability to choose subordinates, consistently finding men with abilities almost the equal of his own, then trusting them to work by themselves. He was cautious when it was right to be cautious, always understanding the real extent of his power. He is almost unique among great men in that he seems never once to have overreached himself. He planned each battle and each aggression with great care, studying the problem and making his preparations with ordered deliberation.

We can get an idea of the care and thoroughness with

which Genghis planned from the account of one of his later campaigns. He reasoned that to beat his chosen enemy he would require an army of ninety thousand men. And, to make their mobility certain, each should be provided with five horses. They must have provisions for six months, which meant a baggage train of five thousand camels. Such were the discipline and method of the remarkable system he had built that all this was collected together at the appointed place and time from the resources of a nomadic people. When a man of this method and genius took up the role of warrior chieftain, the battles started to become decisive. Not only did you lose if you went against Genghis, but he was good to serve under, whether you went as a volunteer or came before him as a beaten foe. More and more tribes and peoples gave him their allegiance, until at last the powerful flowed together under his leadership like drops of mercury running to the bottom of a bowl. They elected him king, or Khan, and gave him the name, Genghis, which is said to have been that of one of their gods.

Genghis Khan now had to make war in order to keep his followers happy, and the only enemies left were the civilized states bordering the steppe. He turned his genius to fashioning an army capable of beating the regular troops of the Chinese empire and other powerful settled states. Plunder, adventure and new land, were to be the gain.

The chosen enemies could all field more soldiers than the Mongols, and yet they must be beaten decisively and terribly, so that they could be made to hand over their possessions. It would not be good enough to have pyrrhic victories with heavy nomad losses, for a dead man cannot enjoy plunder. Genghis Khan set out to organize his army so that it knew, not only that it would win, but how it would win.

A Mongol army in battle array was drawn up in squadrons, each of fifty or a hundred men, each in five carefully lined ranks. The first two ranks were men in the scale armor, with steel helmets, and carrying lances; even their horses were sometimes armored with a suit of iron scales. The rear three ranks were of bowmen, clad in leather only, and with a large supply of arrows of several types, among which were heavy shafts with armor-piercing heads.

The squadrons were all properly drilled, under unified command, and trained to respond to waving flags and flashing lights. Their general oversaw the battle from a vantage point from which he was in communication with his fighting squadrons and their reserves of horses and men. Each squadron could approach the enemy in a way reminiscent of a cohort of a Roman legion, with its checkerboard of men melting between each other to discharge their missiles and then retire. The squadrons rode up to the enemy, the armored front ranks parted, the horse-archers rode through to loose storms of arrows. If the enemy charged, the archers melted back through the lines of armored lancers, shooting backwards from the saddle as they went. Often a counterattacking enemy would be led to pursue fleeing archers until it was fatally far from friendly lines when the fleeing horsemen became a closing circle, and swishing missiles withered the lost force away to nothing.

The Mongol squadrons went on shuffling their horse-archers through their ranks of armored cavalry, incessantly shooting down the men standing against them, feeding extra squadrons to the ends of their lines until the enemy was in an envelope which launched endless volleys of the swishing shafts about them until the ranks thinned, and the hopelessness of standing there ate into the resolution of the bravest hearts. Then their array would crumble and the Mongol heavy cavalry would charge to finish the slaughter the archers had begun. Army after army went down to smaller numbers of Mongol horsemen in this manner. There was no tactical answer to a Mongol army of horsed archers in the open for they had superior fire power which could not be met. A horse archer was the product of a lifetime's training, and only the nomad way of life provided this.

Some generals tried to frustrate the Mongol tactics by fortifying the battlefield. They planted rows of lances in the ground to keep the Mongol horses out of bow shot, hoping to lure them into attacking on foot, or so frustrate the Mongol advance that they could secure tactical surprise for a counterthrust of their own cavalry. But the cautious Genghis Khan and the other Mongol captains would never be drawn. They merely waited until thirst and hunger drove the defending army to move, and then they attacked it, for the Mongols had not only superior fire power but superior mobility also. They could not be frustrated by fortifications in the field, as long as they could cut the supply lines to those fortifications. Indeed, there was no way of beating a Mongol army in the open. Their technical superiority was absolute, and remained so until the invention of hand-held firearms let the opposing soldiers be as well armed as they.

In forested or rough country, the advantage of horse and bow was not so strong and skillful generals could sometimes prevail against them. They rarely did so, however, because the Mongol generals were ever prudent and would always refuse battle except on ground of their own choosing.

The hopelessness of trying to stand your ground before one of Genghis Khan's armies became gradually apparent to the military men of the nations they attacked. There seemed to be only one possible way of resisting them, and this was to shut yourself up in a well-provisioned fortress and wait until the Mongols went away. This, indeed, was the only military answer before the invention of explosive small arms, and it sometimes worked. Some of the fortified towns of China managed to resist until the horsed army moved on in quest of softer prey, and others were able to buy off the army of plunderers. But aggressive war had now become part of the Mongol way of life. Like so many other peoples before them, the Mongols, under a great captain, had forged an unbeatable instrument of war and then were carried on by the momentum of its success to demand conquest after conquest.

The military genius which had perfected discipline in the ranks of horsemen now studied how to take armed fortresses. Genghis Khan and his men would never achieve the absolute technical superiority in siege warfare which the horse-archer gave them in the open, but they made themselves masters of the contemporary art. Every siege engine known to their times found its way into the Mongol baggage trains, and the nomad horsemen of the steppes, under the fierce discipline they had accepted, became masterful engineers in their use. It is sometimes said that they even had cannon; certainly they learned to bombard and take virtually any town they thought they needed. They plundered their way, not only through all the border towns which were their original target, but, as their ambitions grew, on and on until they placed some of their descendants on the imperial thrones of Asia.

The nomad armies were never beaten. In the end they merely faded away. Like the decline of more conventional empires, this has often been seen by moralists as the result of a loss of spiritual purpose in the descendants of the conquerors. They grow soft, take to loose living, wallow in their harems, and forget that the martial graces are supposed to be superior. But such moralizing is not necessary to explain the ebb of the Mongol aggressions. The need, which had caused the people to throw up that dreadful army, had been satisfied. The wars had first diverted the frustrations of a rather crowded people with adventure and plunder, and then had removed the cause of those frustrations entirely by effecting an armed emigration.

The wars won new land for the people, and the people moved en masse to occupy them. The steppes were no longer crowded; nomads could live there as nomads always wanted to live; with a little fighting now and then to keep the culture going, but without the pressures that led them to the savagery of Genghis Khan's boyhood. There was no more need for iron discipline to support an aggressive war, and the people preferred their freedom. So the force from the steppes was slackened and the armies faded away. The steppes would be quiet for centuries until another accident of the rains put too many people there again; or until the invention of firearms, tanks, and the Union of Soviet Socialist Republics, ended the cycle of nomad aggressions forever.

CHAPTER SEVEN

REBELLION IN AMERICA

"Antagonism between citizens and soldiery flared up in the so-called 'Boston Massacre' of 5 March 1770. A snowballing of the red-coats degenerated into a mob attack, someone gave the order to fire, and four Bostonians lay dead in the snow. Although the provocation came from civilians, radicals such as Samuel Adams and Joseph Warren seized upon the 'massacre' for purposes of propaganda. Captain Preston and the British soldiers were defended by young John Adams and Josiah Quincy, and acquitted of the charge of murder brought against them; but the royal governor was forced to remove the garrison from the town to the castle, and the strategic advantage lay with the radicals."
—S. E. MORISON and H. S. COMMAGER
The Growth of the American Republic

Four hundred years ago the land which is now owned by the United States of America was already full of people; indeed, it was crowded with people. Yet they were very thinly spread on the ground, all the same. Their niches made certain that this was so.

The Amerind peoples hunted, gathered and grew a little corn. They made everything they needed, from clothes and weapons to agricultural implements, with tools made entirely from chipped stone, bones or wood. They were without metal. The only domestic animals they had were dogs. The fastest possible way to travel was to run, or to paddle a canoe. The people made houses of skins or branches or sun-dried mud; depending on which part of the continent they lived in.

But though the niches of the original Americans made them rare it is certain that they had quite filled the land. They had lived there for ten thousand years and multiplied to press upon each other's lives. It is proper to say that they were "crowded." We can, in fact, find strong evidence for this crowding in the last records of these people; tribal war was a way of life with them and it has been possible to trace out the borders of some of the tribal lands for which the people fought.

But this ancient homeland of the Amerinds was discovered by people of quite different culture and experience, equipped with a different technology and trained to the niches of agriculture and city living. These new people would not be crowded until they were ten times, or a hundred times, as numerous as the Amerinds. To them the land was empty; and so they took it. They made it yield human living on the scale possible where the tools are made of steel. They came as colonists, bringing their culture and their ambitions for niche-space with them, just as Greek and Roman colonists had settled other barbarian lands two thousand years before.

Then followed the short history of modern America.

The North Americans have gone through their cycle far more rapidly than did the ancient Mediterranean states, but the main stages in the process follow the old pattern clearly enough. There were many small colonies from different parent states, speaking different languages, yet sharing a common technology. The French, Dutch and English colonies fought one another as they became jealous of each other's trade, or as they sought to preempt land for their own use. They alternately bullied, or made allies of, the barbarians who were the original owners of the land and they used the barbarians as soldiers in their own wars. Eventually one group of colonists in North America, those who spoke English, came to dominate the rest.

And when the colonial struggles were resolved, the European settlers went on to take the whole continent from the Amerinds by the most successful war of aggression of which we have record. As in all successful aggressions the victory was won by superior weaponry and training to war. It was firearms against bows and arrows, a disparity that makes the superiority of a legion over a mob of unarmored swordsmen seem trivial. It was so easy we think it hard to call it a war of aggression at all. Yet it was.

The initial building of an English-speaking empire in America was followed by the expected rapid rise in the population, a process which still goes on. This cycle of history is not yet over. But some of the symptoms that come late in a cycle already are showing: spreading bureaucracy, restraint on individual freedom to do as you please, welfare and cheap entertainment for people who must live in large cities, inflation, and wage and price controls. The history of the United States also tells us of changes in the human niche that lead to revolution and of how liberty itself affects breeding strategy and the way we learn to live.

In the summer of 1970 soldiers were ordered into many American universities. I saw them march to occupy the campus where I profess, The Ohio State University in Columbus. They came equipped as for war, with steel helmets, loaded automatic rifles and fixed bayonets. They found not the welcome of liberators, but the angry hostility of several thousands of our students. Soon there were the scenes made familiar to everyone by television. The clouds of drifting white gas against the green of the college lawns, the ragged lines of khaki-clad men with faces made ugly by respirators, the colorful crowds of shaggy students milling away from the tear gas, jeering, with an occasional daring one hurling back a smoking canister to win an approving chorus from the crowd. Over our once sleepy lecture halls helicopters clattered their infernal din, so that neither the voices of reason, nor of folly, could be heard. Television crewmen made their slow way across our beautiful central lawn, which we call "The Oval," trailing commotion with them, as students threw rocks to provoke gas bombs, making sure the TV crews had something to shoot. It was like a war scene on a Hollywood set, with clouds of smoke, explosions, lots of noise, lots of crowd-scene extras, and the camera crews busily at work, the whole war looking both amateur and sham.

But in the days that followed that first excess the affair settled to something more serious. The soldiers were now posted as a line in front of the administration building, the hot Midwestern sun glowing down on their helmeted heads, their rifles pointed in a threatening gesture that must have been a wearying pose to hold. Six to ten feet in front of them was the edge of a herd of students, always several hundred strong, disputatious, taunting, and becoming increasingly bitter.

The faces on either side of the bayonets showed no comprehension of each other's purpose. There were all the expressions which you would expect at such exciting doings outside the experience of everyone there; expressions from exhilaration and triumph through hostility and fear. But the unifying theme was this lack of comprehension of each side for the other. Both knew that they were in the right.

The soldiers were there to protect American liberties, which were being weakened by some strange, subversive thing. They obeyed the orders of mature, elected government. Their duty was clear. The students had come there by a more indirect road. They had first felt, along with many of their generation, that something was wrong, then at the university this worry was sharpened, as the cleverest of a generation were collected together where they could talk. Their university was very large, as large in fact as many of the armies on which the fortunes of nations of the past had sometimes depended, forty thousand people of military age. But the university was not run by the sorts of people who lead armies in times of trouble; people alive to morale and the human factor in all their thoughts, but by bureaucrats operating as if in the business of commerce. The honest labors of these businessmen provided for the students' buildings, money and things; the physical material which a modern university must have. But the bureaucrats did not understand, and could not administer to, that worry of a generation which was growing articulate among the libraries and laboratories provided.

There was, of course, the frenzy of Vietnam in the air, and without that frenzy there might have been no riot. But at Ohio State the riots began by a calculated insult to authority, which had no obvious connection with the war. On one weekday, a group, bent on trouble, closed a university gate that was usually closed only on football Saturdays. The act was popular because the shut gate stopped a river of traffic from flowing through the campus. We had all long wanted that road closed. The act, therefore, had wide appeal to those of rebellious spirit, because it asserted that cars could be kept off a campus at times of study as well as when we lent the place to popular amusement provided by our football team. And the act was as calculated a snub to university authority as could be devised. It produced the expected response, what those who had shut the gate surely wanted—buckshot, tear gas, the uniformed panoply of force.

Many of the later provocations by student groups were of the antiwar type, involving the usual marches on buildings where military studies were taught. An antiwar march could always pull in a small crowd in that generation; but antiwar had little to do with the crowds of up to ten thousand who growled on the Oval. As a focus for violence a black activist group calling itself Afro Am was more effective than the draft resisters. In truth there was a sea of grumbles for agitators to exploit.

Closing the gate on Neil Avenue was just a start. The response of sending a sheriff to open the gate could at once be used as evidence that administrators were "insensitive." Political-minded students simplified the general disquiet of their peers into complaints against those university officials who gave them buildings but could

216

not help them to peace of mind. Folly and good intentions then walked together, through provocation and counterprovocation, until the businessmen knew their helplessness and called upon the state. But all the state could offer was armed men. So the soldiers came.

There is probably no university on earth, outside those in places kept down by tyranny, in which the tramp of soldiers come to tell students what to do can fail to rally most of them to the ranks of the most radical. This was the explanation of the exhilarating battle scenes of the first day. In the later days, when the soldiers were drawn back to a guarding line, the exhilaration was gone and only the politically conscious were left to face them. But these few knew that they were right. The disquiet of their generation led them to want to make changes in the accepted order. Soldiers had come to make them conform; they would not conform; they would face the soldiers in the name of American liberty.

As I watched those opposing self-righteous lines the feeling grew upon me that something very like this confrontation had been seen in America before. Two hundred years ago, in Boston, an indignant herd of young men had faced a line of British soldiers, who threatened them with loaded firearms and fixed bayonets. That time, too, both sides had been sure they were right. The soldiers were there to keep the peace, to support the traditional and orderly way of doing things, to preserve freedom under the law. They obeyed the orders of mature, constitutional government. Their duty was clear.

And the British soldiers in Boston had the support, not only of an obstinate king in a faraway country, but of many of the upright citizens of the thirteen colonies;

well-meaning people who thought that good intentions were enough. They were not an isolated few who thought this way, for it is said that, in the war which was to follow, the State of New York sent more volunteers for the armies of King George than it did to the revolutionary armies of Washington. It was the older, wiser heads who counseled moderation in respecting the soldiers' duty to keep the peace.

But the Boston crowd was of younger men, bred in a generation of disquiet, the inarticulate messengers of a new way. The crowd shouted "Liberty" in the faces of the dutiful defenders of law and order. As in Columbus two hundred years later, the old way faced the new way; orthodox wisdom faced impatience with its consequences; the generation of maturity faced a generation which was bewilderingly different. In Boston the guns of the outnumbered and frightened soldiers went off, sounding those celebrated shots which were heard round the world. Even as I walked among our students, watching the frightened faces of the soldiers in Columbus, the rifles of a sister regiment were going off to kill four students of a sister university at Kent, a few tens of miles away. The parallel was so close that it seemed proper to look for some common pattern in the causes also.

By any reading of those two confrontations, two hundred years apart, there were issues of ways of life involved. Each time an old order faced a new. The niches of the people were changing, and it ought to be possible to describe the happenings which led up to each confrontation in terms of changing numbers, changing patterns of breeding and resource, and the changing aspirations which the new densities of the people made necessary.

The Boston scene had its beginning two hundred years earlier still, in the England that had adopted emigration to America as her national policy. The American colonial niche was at first an English niche, fitted for life in contemporary England. Even when training for the English niche began to be lost by those born and bred in America of English parents, colonial life was still molded by the steady replenishment of those trained in England. And right up to the fateful days in Boston the governors of the colonies came from England and were bred in English ways. To know why the revolutionary war was fought it is necessary to understand the niches of the England of those times.

That England was an island state, densely populated, but with both numbers and opportunities for life expanding. The English had turned their island into a productive garden, with little wildness left. Agriculture, industry and commerce were each collecting into large-scale enterprises. It was a society in which the younger sons of the better-off were already taking to trade and foreign adventure, achieving broad niches by the use of other people's lands. And the people had fashioned rules of law and government suited to their tightly organized garden of an island.

The niches in which these English people lived reflected the organization of their developed state. The English who stayed at home expected their futures to be narrowly circumscribed. There was a place for each in society, but little choice of place. It was a society of castes, liberal by the standards of many a caste system, but a society of master and man nevertheless.

Yet the English were thriving on the new developments in agriculture and industry. The optimum family size was large for all classes and the population tended to grow rapidly. This rise in the numbers of the English is well documented. The English themselves knew what was happening, and they were alarmed lest their prosperity might be undermined thereby. Contemporary writers set down the dilemma and one of them, Richard Eburne, showed that salvation was to be found in the mass export of people to colonies.

In *A Plain Pathway to Plantations* (1624), Eburne sets down the predicament and offers the American solution. There were so many people that they could not be properly housed; the landlords were demanding exorbitant rents; the poor were oppressed; the desperate were putting up shacks wherever they could; the countryside was afflicted by rebellious people. There was no law and order in the city streets. And the remedy suggested for this state of affairs was "send our surplus young people to America for their own, and their country's good."

Eburne's book argues both the trouble and the remedy for 150 pages of quaint but compelling prose, a flavor of which is given by the following quotation:

Truly, 'tis a thing almost incredible to relate, and intolerable to behold, what a number in every town and city, yea, in every parish and village, do abound, which for want of commodious and ordinary places to dwell in do build up cottages by the highwayside and thrust their heads into every corner, to the grievous overcharging of the places of their abode for the present and to the very ruin of the whole land within a while if it be not looked into, which if they were transported into other regions might both richly increase their own estates and notably ease and disburden ours.

Eburne's book, with its reasoned answer to almost every ill of contemporary society, was very widely read by the influential men of England. Its program became, in effect, national policy. And through that policy the English undertook to take America and make it theirs.

By their American conquest the English in England had made safe, for centuries, their ancestral way of life. The friction between landlord and tenant, between rulers and led, between rich and poor, of which the *Plain Pathway* spoke, at least got no worse. And by reason of the hope which the new room, the new carrying capacity, gave to every man and woman that their children's lot might be better than their own, the extraordinary conquest actually improved the peace of mind of all. Pressures toward caste and rank were stayed and the English concept of freedom under the law grew ever stronger.

The English in America brought with them that English concept of freedom under the law, complete with deference to authority and restrictions on behavior appropriate to life in a full-up island laid out like a garden. American elders read of the island life, some were homesick for it, and others learned about it from the regular flow of immigrants who brought it with them. But the way of life, the niche, found to be so satisfactory for England was not so appropriate to Englishmen in America. Why accept the social system which had been found necessary for a well-peopled garden? Here there was room for everyone to do as they pleased, respecting their neighbor perhaps, but not neatly fitted into a neighbor's affairs as the English in England were. Why be master and man when there was room for everybody to be master? A new form of liberty was possible in which each could pursue happiness with much less deference to vested interest. Gradually the niche of an American took shape, described though it was in the language of England.

As the numbers of Americans grew from the first few hundred to a million, and then three million within the scant generations of less than two centuries, the English in America were less and less beholden to supplies from the garden factory across the ocean which had sent them out. As the need for things English waned, so the propaganda for the English niche waned also. For more and more Americans the English way was no longer a proper way, and young people began to grow up in puzzled and unhappy awareness that what their elders and betters did, with the idea of promoting liberty and freedom under the law, was somehow not right. They argued and questioned and took counsel together. The end of it was the night in Boston when the soldiers shot, and the war between Englishmen which decided that those of them in America should henceforth live in a new niche divorced in bitterness from the old.

English governors had brought the war down on their heads by well-intentioned bungling, like the bungling of fair-minded administrators who would, one day, induce the students of an Ohio university to howl their resentment at American soldiers. They meant well. They thought themselves eminently fair; preserving order, protecting the rights of others. Even the notorious folly of the Stamp Act was no more than simple justice, a way of paying the British army and the British fleet for services rendered in pushing the Indians and the French away.

It was fair-minded reasonableness that sat in the minds of the men who told the soldiers to control that Boston crowd, and the reasonableness of their position was self-evident to mature thinking people. But other things were self-evident to the crowd. A new niche had been found which gave wholly new possibilities for the people in this still-empty land. All that needed to be done was to clear out the old way, with its irrelevant obligations. Beat in battle those who preserved the old things, and nothing would obstruct the hopes held out by this new niche for as far into the future as people could see.

Niche faced niche across the British muskets in Boston; a niche suited for a crowded island and a niche suited to the almost unlimited possibilities of a new continent. Both peoples used the words "freedom" and "liberty" when they described their purpose, and, hence, the niches they were to preserve. But the British talked of the regulated freedom of a complex society wherein choice was necessarily limited by the numbers of other people needing their shares, and the Americans talked of the much wider freedom possible when opportunity and resource were virtually unlimited.

The guns went off, and they went to war, and the new idea won.

Like the more blatant aggressions, revolutionary wars are fights over niche-space but the quarrel is over what a nation already owns instead of over other people's land. The aggressor has a new idea and the victim clings to an older way of life. The spoils of war are the right to use the national resources for the winning niche. As in other aggressions, those who make the war and fight the hardest are those with most to gain, the classes with expanding ambitions, the people in the middle ranks of society. Karl Marx denied this, but history is not on his side. The several revolutions of England—Magna Carta, Parliament against the Stuart crown, William of Orange and the Whig oligarchy—were pushed through by newer affluence against established power. The French Revolution came only after a broadening middle class was without power before an entrenched aristocracy and it was this new class of comparative affluence who came to the top in the Napoleonic empire following the revolution. Even in Russia of 1917, the prime agitation was from bourgeois intellectuals, though the desperate mass they led saw and wanted the better living that should be possible from industrialism. Russian serfdom was tolerable in former times but not when mass-production techniques made a broader niche possible.

The ecological hypothesis says that all aggressions come when a society must find more resources to meet rising ambitions, and the aggressions of revolution conform to this rule. When a society is ambitious for broader niches we expect families to grow larger so that the population grows and again we find rebellion in America, France and Russia at times when the numbers grow. And there are very clear signs that new niches are being learned in the years before each great revolt, sometimes through deliberate education. In America the learning came from combining English rhetoric about liberty with the experience of living in a virgin continent, and its results are made clear enough in the cry "Every man his own master." Finally the ecological hypothesis predicts that victory can be won only through superior technique. This requirement may be lessened a little for revolutions, since the wars are "civil," because the losers need not surrender to a foreign foe but merely agree to let their neighbors govern. Even so, as the American War of Independence shows, final victory comes only when the rebels find a winning military formula.

Wilderness fighting decided the War of Independence. The way to win battles in the wilderness had been learned twenty years before the revolutionary fights, in the Indian wars, and the sternest lesson was when two regiments of British infantry were destroyed in the forest by half their number of French irregulars and their Indian allies, all armed with rifles. The British were marching on Fort Duquesne, under an experienced, competent and brave commander, General Braddock. In the British ranks were a Lieutenant Colonel G. Washington and, serving as wagon drivers, two young cousins, Daniel Boone and Daniel Morgan.

The expedition set out from the Potomac River to cover 120 miles of forested wilderness, crossing the Allegheny mountain range as they went. It was summer, the great trees of the forest heavily in leaf, the long line of red-coated soldiers and their wagons winding through the gloom of the forest floor like a worm through long grass. The soldiers were armed with muskets, and they were very well trained in their use, an army of professionals, tough, secure. They could line in ranks to fire volleys at a word of command, they could form squares, when they became a death-dealing fortress, sweeping the ground in front free of horse or man with precise salvos of shot. They could reload with calm precision, in unison, to given commands, stepping back to let their rear ranks pass between them to fire. They were the missile men of a Roman legion, though armed with muskets instead of javelins. But now they were in a great forest, snaking in line between the trees.

Two thousand years earlier three Roman legions had wound through a similar forest in Germany, the legions under Varus, hunting Arminius and his German barbarians. The legions had had their baggage wagons with them too, and they also had pressed on into the wilderness for day after day. But then gaps between the trees had filled with bowmen and men throwing spears, and

225

the legionaries had been picked off one by one, until they were all gone and the defeat was so bitter that Rome abandoned the conquest of Germany for good. Braddock's fine soldiers were treated in the same way, also by barbarians lost between the trees, but using the first rifles and long hunting guns instead of bows and javelins.

The shots came from out of the woods, the soldiers knew not where, and they were accurate shots. The soldiers formed their lines as best they could and fired their salvos into the gloom where the French and barbarian marksmen hid, but they had no target at which to aim. Every man had a neighbor smashed beside him by mushrooming lead balls striking out from the forest; there was the stickiness of blood, the sound of agony and the smell of mortal fear; but scarcely an enemy to be seen. Training helped for a while; they closed the gaps and followed their officers in dashes at the spurts of fire between the trees. But the officers were shot down, and shot down again as fast as they were replaced. General Braddock had five horses shot under him and then took into his lung the bullet that killed him. Colonel Washington had the skirt of his jacket torn through by bullets, and his horse killed. Of the 87 officers, only 26 survived; more than half the soldiers were killed or wounded. Two hundred French and six hundred Indians did this to 1,460 of the best-trained soldiers in the world.

The revolutionary army used this technique against the British and the German mercenaries brought over to serve England's Germanic king. There was not such another ambush in the wilderness, but the land of the thirteen colonies still had plenty of forests and woods. American frontiersmen knew how to use rifles, and the memories of the French wars let the local commanders know what sharpshooters might achieve. The Continental Congress was busy raising companies of riflemen even before they got around to appointing Washington to command the army.

The British side knew about the value of riflemen too, and there were enthusiasts for the new arm in the British army in America. It was no unique trick that the Americans had, but it was a trick that particularly suited the American cause. In open country and near the main armies, the rifles of the day were so slow to load that they were a skirmisher's weapon only. There could be no massed volleys to smash down an opposing formation and rifle companies turned and ran when pursued by disciplined lines of musketeers. They had to. But there were many woods about in which to hide and if there was some support from regular infantry, and your own woods in which to retreat, then the riflemen could turn the tide of battle. They did this for the American armies defending their lands, but they were much less suited to the punitive operations of the British expeditions. And they turned the tide at several notable fights.

At Saratoga, riflemen were decisive. This was the fight in which a major British army under General Burgoyne was so badly beaten that the British cause was seen to be lost. And the critical stroke that decided the issue was the crumbling of the British flanks by riflemen coming through the trees. In front, the British position was open, bare fields through which the disciplined infantry could deploy in their disciplined way; affording a free field of fire for the British cannon. The British flanks rested on blocks of woodland, a good textbook position because woods stopped any regular body of troops from forming against you. But Daniel Morgan, with his experience of defeat with Braddock, wended through one wood with a company of riflemen, and a second company wended through the wood on the other flank. They shot their way into the disciplined British lines, forcing the retreat that led to surrender.

After Saratoga the end was certain, though England's king was stubborn and his country had to lose more wealth and friends before he would let the colonists go. American technique in battle was now good enough to make sure that the rebellion could not be crushed, however much the English copied the rebel methods and tried to use the wilderness themselves. The colonists had the special advantage of all revolutionary armies that they fought in their own country, where they could find succor, escape and a chance to regroup. Revolutionaries are always, as Mao Tse-tung put it, fish with a sea to swim in. Their military advantage need not be so devastating as is needed by an army taking other people's land, as indeed it can never be because the rebels fight people of their own culture and military experience. But they must have something like parity in weapons and then must develop advanced tactics which will let them survive to go on swimming in their sea. The Americans put together rifles, the forest, and great distance into a combination which, in the long run, could not lose. The aggression of people with new ideas of a proper niche succeeded in its fights so that the resources of America were used to support the new ways instead of the ways of Old England.

The fight of the youthful colonists with the armies of their erstwhile king did for the Americans what aggressive wars did for expanding peoples of other emerging empires. They had learned competence in contemporary war and a keen understanding of the advantages given them by geography. They knew they would be able to cope with any force likely to be sent against them on the continent they claimed. Now the ambitions of the people could grow almost without check. They could proclaim their new niche, with its wide freedoms for individual choice, and see that the rising aspirations they had unleashed were met by taking over more and more of the land which had fallen into their power. European governments who still had claims against them by reason of their original aggressions against the Indians were pushed aside, beaten, or bought off; and the boundaries of the state were thrust, like the empires of old, to the natural boundaries of rock and sea.

Revolutionary armies always turn against their neighbors when they have secured the victory at home, unless their neighbors are exceptionally powerful. Napoleon's armies rolled out of revolutionary France shortly after the process of lopping off heads was complete. The Chinese Communists invaded Tibet just as soon as their victory in civil war was secure, and they tried many a nibble at the land of other neighbors as well. The Red Army of the infant Soviet state went out to conquer as soon as its leaders dared—Finland, Poland, the little Baltic nations, and Mongolia were struck even before Hitler's war—and the Red Army has not weakened yet. Even the ancient triumph of Macedon over Greece was victory in civil war that had much in common with modern revolutions, and the triumphant soldiers of Macedon then stabbed their way through the Persian Empire. From revolution to conquest—it is a common theme in history and, like so much else about war, fully predictable by the ecological hypothesis. Rebels go to war that they may win the resources for an ambitious niche, as do the fighters in any other aggression. When the first victory comes and they are masters of the state, they have learned that aggression pays, and so they try more of it.

The American empire was nearly unique in the history of the world. It was made of good, almost virgin land. Adventurous New Englanders found themselves with a capital gain of enormous size, a gambler's jackpot the like of which would never be seen again. It meant an almost unimaginably large number of broad niche-spaces, with hardly any of them filled.

Moreover, the American civilization knew techniques which were themselves sufficient to extract very many high-quality niches, even without the glut of untouched land. The number of people who could be supported in affluence was, therefore, almost unthinkably large. Possibly more important still was the fact that there could be almost unlimited choice for the few people who were there to exploit the new resources. For choice becomes possible when there are surplus opportunities from which to choose: I choose to follow this vocation, and I therefore leave that vocation vacant.

In old England there was little freedom of choice because the garden was full; people had to be coerced by social pressure into their appointed ways of life. But in New England this coercion was less necessary. There were vacant niche-spaces to every hand. It really was true that there could be freedom of choice. Even though imperfections in the social mechanism might give some citizens rather less choice than others, absolute freedom to do as you please was near enough a practical proposition for people to believe in the possibility and to try to make it real. American society then developed round the idea that all the individual people should be able to develop the most that was in them, to achieve their utmost in whatever walk of life they had been best endowed by their maker, without arbitrary constraint of government or their fellows. It was the greatest dream that people ever dreamed.

In these circumstances the human breeding strategy is expected to produce large families, and the record shows that it did. Large families became an American way of life. We see this in the prevailing Protestant ethic of the times in which the rewards of success are the happy marriage and the flourishing family of numerous offspring. The offspring, moreover, are expected to be able to better themselves, showing that society expected to produce the resources not just to feed these large clutches of children, but also to lavish on each the training and resources needed to let it live in a large niche. This was all part of the dream; for those generations a perfectly possible dream.

Yet the Americans attributed their good fortune, not to their unprecedented blessing of excess resource, but to the moral worth of their American ways. They throve in their new freedom, where every person had the same chance of fulfillment, where no necessities of social order decreed that some people always had less liberty than others; and they called to others to join them in America, where all would thrive together. Even as American numbers grew, Europe was so crowding itself that its governing classes began to press ever more strongly on the mass in their efforts to preserve their own traditional ways of life. As Europe filled with more and more poor people, the hopes they could hold for their children in their native lands were less and less. To them the Americans said, "Come here, forswear those potentates who grind you down, and come and live as we do."

This breeding effort of the Europeans which America

was set to absorb was itself certainly driven in part by the presence of the new America. Cheap American food fed babies in European slums, so that the future immigrants were actually living in the American spaces even before they crossed the Atlantic. More importantly, the knowledge of the American presence gave hope that the children of large families would find niches in the next generation. This American reason for hope was thus added to the possibilities inherent in the new industrial techniques of Europe and was probably a prime cause of the rapid growth of the European population. And when that growing population brought on itself the ills of social oppression and poverty, there was the young United States with its, "Send me your tired, your poor, your huddled masses, struggling to be free."

With immigration and a sustained high birth rate combined, the population of the United States rose from three million to two hundred million in a mere two hundred years. The population was then more than large enough to fill even the immense American space with people living as those who declared the American version of liberty had wanted to live. Fortunately, there had occurred meanwhile those extraordinary technical advances in manufacture and commerce which have let the modern Western world multiply its carrying capacity by many times. The ingenuity of American workmen had added more niche-spaces almost as fast as they were filled by immigrants and babies. For long it seemed as if this could go on forever, that liberty was a self-fulfilling process which would produce more resources and more room for liberty, as long as you cared to go on working at it. More people meant more needs, which let the lives of still more people be satisfied in meeting those needs.

But not all the resources of the American niche could

be provided by technical ingenuity, for many needed space. The essential American individualism itself implied elbowroom, so that each could go his own way without too much harm to his neighbor. As the numbers grew, the social organizations of more crowded people began to appear. The rich began to pass on their riches to their descendants, and the poor their poverty, just as they had done in the Europe of the potentates for whom the Americans had such contempt. Liberty was now something at which all the people did not get an equal chance, and a fundamental tenet of the American way of life was denied.

This change in the American reality worked its way into the social consciousness only slowly. For a long time it could be truly held that there was still enough room for all, that with the technical ingenuity at which Americans were preeminent, her resources could be so multiplied and distributed that this unequal sharing of liberty could be put right. It was in this spirit that the social programs of the New Deal period and the postwar administrations were undertaken. For a time they seemed to totter toward success, as the numbers of actually hungry, workless and hopeless slumped away before the discipline and purpose of technical resolve. But the labor was bound to be vain in the end, because its goals were impossible. Society was trying to grant a niche of untrammeled individuality to all the citizens of a space which, for that purpose, was already full and when the numbers of those citizens were constantly allowed to grow. The American empire was well on in its "age of the Antonines," with good men governing, with public works for the benefits of the people. Large-scale farming and manufacturing drove the people from the countrysides to the towns, where the governors were forced to give them a surrogate for their promised niche. To not a few of them the surrogate came in the form of a welfare check. It was in this America that, in the direct parallel to the Boston Massacre, herds of unruly civilians thought it necessary to taunt lines of soldiers over the muzzles of their guns.

In the 1770s the mass of the people knew that the old way of doing things would not suit the circumstances of their hopes or their times, however good and well-intentioned those old ways were. Niche faced niche across the muskets of the British soldiers, the older uncomprehending the needs of the new. But in the two hundred years that have passed what was new is now old; a generation has been reared in suburbs, sated with material surrogates that are not adequate substitutes for the self-expressing freedoms of the past. And a parallel generation, much of it black, has been raised in cities without even the material substitutes for free living. This is the simple and inevitable consequence of allowing the population to go on growing.

The thinking young of 1970 knew that something was wrong; that the old ways had failed in some mysterious way; that the traditional rhetoric of untrammeled freedom did not fit the world as they saw it. Indeed, they saw that the traditional right of everyone to do as they pleased was inappropriate to the society in which they lived. For their society was one of dense settlements, with the vacant niche-spaces rapidly filling, when the choice of a job or a vocation in life was very far from wide open. The young were adapting to an America which was as tame and full-up a garden as England had been in the days of first settlement. They lived in a new American niche, as different from the niche of the generations that went West as the revolutionary niche had

234

been to that of old England. But America was still governed in the rhetoric of a constitution framed in that earlier time.

The generation gap was a separation of niches, the old clinging to definitions of life style now outdated, the young adapting life style to the new densities of the people. In the universities of 1970 and 1971, there were blunders in the handling of people not unworthy of the cabinet of George III itself. Our administrators had meant so well, they were fair-minded, hard-working and honest, but they could not understand that they had to treat with people who were pursuing a new idea. For a few days a new niche stared at the old across the muzzles of M.1 carbines. On both sides of the guns the purpose was to defend "liberty," just as in old Boston. But the new density of people had caused "liberty" to have changed its meaning yet again.

CHAPTER EIGHT

THE STRUGGLE FOR EUROPE

BY THE year 1600 an entirely new chapter in the history of the human kind had started in Europe. The essentials of this new Western way were technical ingenuity and a general literacy or communion between countries which let people copy skills from one another as fast as they were invented. Soon this way would let a handful of generations tap all the energy stored as coal and oil in the earth and increase the number of potential human niche-spaces almost beyond imagining.

In 1600 the new ways were still just beginning, but even then they were further advanced than we sometimes think. Wind and running water were used to drive engines for pumps and mills. Industries were concentrated round them. Foundries used coal to smelt iron and made things inconceivable to the civilizations of ancient Greece or Rome. Armies fought with firearms, and the sea lanes were held by ocean-going ships firing broadsides of cannon. These things are modern enough, but in agriculture we find the people of 1600 more modern still. Farmers in the Netherlands grew crops in a four-year rotation; wheat, clover, grass, turnips, with animal manure and marl being regularly returned to the land. Improvements in crop husbandry since then have been mere embellishments. In Tudor England, land was already being enclosed for the intensive rearing of livestock, and to promote proper crop rotation. By 1544 Dutch engineers were being imported to England to drain English marshes. Coastal acres of the Baltic Sea were being reclaimed by Sweden. People suffered hardship when driven from ancestral landholdings, but the thriving cities of manufacture were already in being to soak them. Agricultural writers of the times had begun bemoaning a drift from the land which put their labor costs up.

The decisive advance was not in any particular technical skill but was rather in the habit of being technically ingenious itself, a habit that spread throughout the Europe of Christendom. The people shared a common code of moral conduct taught by centuries of allegiance to a common church, which kept the peoples of the several nation-states in close touch with one another. There was a common technology and a common standard of life. When the habit of technical ingenuity set in, all the Europeans got it. All together they could start releasing the resources of a continent. They did so, not as a process of plunder in the manner of those who had subdued great empires by armed force, but each part for the local good. Never before had a common culture developed on a continental expanse the size of a great empire without central direction and central bureaucracy.

In the century that followed 1600 the Europeans thrived with the new agriculture and let their numbers grow. They restructured much of contemporary religion, arguing and fighting about the duties of man to man and of man to God through the length of their continent. They improved their weapons so that, by the end of the century, all fighting soldiers had firearms. They established themselves in the stone-age Americas, and they started their outposts in the ancient and crowded civilizations of Asia.

In the 1700s they populated the Americas to the point of planting small new nations of themselves, and they conquered India, a continental isthmus holding more people than did contemporary Europe, as well as being of a more ancient civilization. Their own numbers multiplied very greatly and wealth spread to a larger part of the apex of the social pyramid. By the end of the century the Europeans were fighting global wars for territory and financial gain. And *then* they invented steam engines. They made use of these steam engines and the other machines that were to follow only in the 1800s and the 1900s.

The use of coal and oil to drive engines multiplied the energy flux that could be used in devising new human niches very greatly and increased the carrying capacity of the earth for civilized people by one or two orders of magnitude. We tend to think of this energy flux and its industries as the overwhelming characteristic of the European West. But a better understanding of our history lies in the realization that the first two centuries of our modern expansion passed without combustion engines. The European nation states had struck out at the world, transformed it, and even occupied large parts of it, before they invented the fossil-fuel economy.

The modern European West has been forged from kingdoms set up by Gothic barbarians after their conquest of Rome. The warrior chieftains founded their own system of government, and it owed nothing to Roman bureaucracy. They built a new social order, as it were, from the ground up. They had no crowded mother city to support from their conquests, but merely took the imperial spaces and parceled out the land among themselves.

Rome's conquerors had no ancient rights in land to respect, and ownership was invested in the fighting kings who led each barbarian band. On this fact depended the very effective constitution under which the people came to be governed, a constitution we know as the feudal system, and which was completely different from the city-based ideas of government of the old empire. The feudal system gave the people a new social contract. In the fullness of time, nearly a thousand years, this new contract did what Roman order never managed and it built societies set to meet rising aspirations through technical ingenuity rather than by herding slaves.

The vitally important principle of the feudal system was that all land belonged to the king and was held in trust by descending lines of vassals. Being a vassal did not mean you were without freedom; rather vassalage was a duty like patriotism, an obligation to the community in which you lived, an honorable state and an honor. Vassals would serve in the armies of their lord with the same mixture of emotions that a modern patriot serves in the army of a nation-state. And a vassal paid annual rent to his lord for lands with the same spirit of grudging willingness with which we pay our taxes.

At the bottom of the feudal social heap were agricul-

THE FATES OF NATIONS

tural laborers who were very closely tied to their immediate lords. These were serfs, or that superior kind of serf called a villein. But they were still not slaves in the sense that the Greeks and the Romans had slaves. They were lowest in a peck order of paying rent, and they could pay their rent only by brute labor in fields or armies, but this is very different from the chattel state of the Roman poor.

Rome of the empire never achieved rapid technical advance, very probably because the wealthy classes depended on the cheap labor of slaves to do their work, and the Roman government was always short of money as a result. But the feudal kingdoms supported their wealthy by rented labor passed up a chain of command. It was a system of coercion mutually agreed upon, at least in part, and it gave incentive for individual ingenuity on a scale quite absent from a slave society. This social habit of paying rents through a hierarchy of entrepreneurial chieftains was, perhaps, the main cause of the technical ingenuity which was to build the modern West.

The feudal kingdoms began their existence with advanced weapons. In the five hundred years that followed Rome's fall, the crudely armored heavy cavalry of the Goths who had conquered at Adrianople had been refined into the armored squadrons of Charlemagne, Roland, Oliver, and the Crusades. But the difference was only of degree, even some of the concepts of knighthood having been present in the old Persian and Roman squadrons who pioneered the armored cavalry alongside the Goths. No other major advances in the techniques of war were made in those centuries, except in building fortresses.

Very many of the lands of the new kingdoms may

have had relatively low populations, particularly if there had been decades of unsuccessful child rearing in the old city-based communities as I have suggested. At least the turn of events was that a society of countrymen was governed by technological soldiers not in the service of a city. This was a beginning as different as could be from the Greek and Roman beginnings, where it was the weapons of people from dense city settlements that imposed their will on the countryside.

The feudal system, therefore, was built by country people under the direction of chiefs equipped with state-of-the-art weapons. Their civilization did not grow out of city-states. Allegiance was through vassalage to a lord and not to a city. The institutionalized poverty of slavery was neither necessary nor suited to such a system, and obligations of vassalage might be expected to promote the social milieu in which experiment with ways of manufacture could be encouraged.

We must not push the argument too far, but it does seem reasonable that feudal society would be a better place for technical entrepreneurship than a slave-owning empire. It was population density, weapons, and the opportunity provided by these for people to broaden their niches which made Christian feudal society possible. And it was in Christian feudalism that the habits of the European West were learned.

The feudal kingdoms had grown into populous nation states by 1600, and in these states the foundations of all our modern technical achievements had been laid. Through advanced agriculture and growing industry it was already possible to provide the resources for ample lives for more and more people. Christendom was enjoying economic growth on a scale probably never known by any society before. This growth is a matter of record, and so also is the inevitable response pushed by the breeding strategy. The numbers of the people in each country rose rapidly.

Also a matter of record are the predictable consequences of rapidly rising numbers in times of hopeful change; the crowding of people into the cities, the drift from the land, farming by businessmen rather than peasants, the actual depopulation of country districts as the total population grew, the taking to trade by people of the wealthy and middle classes, the arming of the traders, and the wars of trade and expansion. For the expanding West, we do not have to argue the changing numbers and the moving of people from country to town as when reconstructing Greek and Roman events. We know they happened.

To the niche-spaces added to each nation by this economic growth was added the very large increment of resources held in the lands of the Americas. This conjunction of new niche-space through economic growth with niche-space from a continent still used by hunter-gatherer societies was unique in the history of our species. It can never be repeated. It yielded the hope of wealthy living on a scale that can fairly be called a glut.

English-speaking people still look back with nostalgia on the days of Elizabeth I, when Shakespeare wrote, when Drake circumnavigated the globe, and when adventure was to be had on every hand. Many of us would cheerfully change places with people of some stature in that society. There was real opportunity then; something properly described as niche-space to spare. There was also capital with which to fuel Elizabethan ventures, a sure sign of earlier ventures brought safely to a finish. All this is predictable as a consequence of a glut of resources. The sense of freedom came from having more niche-spaces than people.

When resources are scarce, then they must be rationed and you do not as you please, but what is convenient for your neighbor. But when there is a surplus, then you can waste some resources and choose others. The broad parameter of the human niche labeled "liberty" requires that the population be low in comparison to the resources for living which it can release, a very rare happening in the human story. But it came about for sixteenth-century Europe. This was the fundamental thing on which a true understanding of the subsequent history of the European West depends.

The constantly recurring chances for liberty brought by the new resources meant that new ways of living were always being invented. The thinkers and the young people were frequently tempted to find new broader freedoms for the human spirit, new ways of conducting affairs, new niches. As the restraints imposed by the physical world grew less, so the restraints of old-fashioned custom seemed irksome and unnecessary. Over and over again, the old confronted the new, with the old resisting in the names of custom and of law and order, the new arrogant in the belief that only they were for "liberty." The American declaration of liberty in revolution was but one example among many of rising hopes leading to those aggressive civil wars which we call revolutions. The last four hundred years of the European West saw a procession of these fights for new freedoms. Most momentous of all, perhaps, was the first, that continent-wide upheaval of the spirit which we call the Reformation. The old Church had grown set in its ways, with codes and dogmas well suited to sustain people in the niches ordained for them by slowly changing societies whose numbers always pressed upon their resource. Now that resources had got ahead, so that there was a glut, there was no need for all people to go their appointed ways. The teachings of Christendom must be modified to reflect this new freedom of choice, and men like Luther did the modifying.

From the time of the Reformation on, the revolutions march down the years in thrilling sequence; the triumph of the English Parliament, the rise of the Dutch Republic, the American War of Independence, the French Revolution, the building of the modern German state, Garibaldi and his tattered thousand, Simón Bolívar, the Communist victory in Russia, and many others which fired the blood and raised the temper in every state of Christendom. All these can be understood from the ecological hypothesis. They were aggressions of new niches against the old, and they capture niche-space with weapons as surely as conquering armies capture the loot of foreign lands.

Throughout this time the people-siphon of the Americas, together with the new ways of agriculture and industry, kept the number of broad niches ahead of the rising number of the people. A measure of this is that wealth per capita grew even as the numbers grew also. Indeed, it was often an open desire to share more fully in the new wealth, particularly by the articulate middle classes, that was so common a quality of the revolutions.

But in spite of the general tendency to revolution, with the increments in liberty for very many, there is no sign of the removal of poverty in the first several centuries of the European expansion. The poor remained poor, and the actual numbers of the poor rose. This is the predictable consequence of the unchecked working of our breeding strategy. As the flux of available food grew, we should expect the poor to be able to raise large families in the niches of poverty. The record suggests that they did.

Numbers, poverty, and total available wealth, all rose together in the centuries of expansion. The process was driven by the continued increase of niche-space in Europe and the steady siphoning off of people and aspirations by the new land in America. More people sent into each generation from ambitiously conceived families did not quite catch up with the expanding opportunities. But the breeding strategy did respond with many more people, even in the first two centuries of growth before modern medicine.

People often say that the huge population bounds of the last centuries came about because sanitation and medicine kept children alive who would otherwise have died. They blame health care for the rising populations; but this is a view not only demonstrably wrong but also in direct conflict with what we know of animal and human breeding strategies. Our populations bounded, peopling the Americas among other excesses, long before physicians could do anything to save babies and often in dreadful states of sanitation. Children died in appalling numbers, but their parents merely had more. The human breeding strategy is nicely provided with the means to replace losses.

In the very short term, medicine and sanitation can make a population jump, but only if the new health care comes to people quite unprepared for it. These parents have begun rearing a larger number than is thought to be optimum by their culture in the sad certainty that some will be lost. When foreign physicians come to save all the children they leave the parents to raise more children than is advised by the culture of their times. This has happened apparently to some poorer communities recently, but the effect should not last long. The next generation of couples will do their reproductive sums in the knowledge that most of their babies will survive, and they will start with fewer. In some poor countries, first exposed to modern medicine after the Second World War, the family size is already adjusting in this way, with a welcome fall in the rate of population growth.

The common view, so widely held by demographers, that high population growth comes from health care implies breeding habits of people which would be quite absurd if applied to any other species of animal that raised a select family in the way that I have called "the large-young gambit." The family size is seen by these demographers as having been set, not to an optimum by natural selection, but by accidents of disease or sanitation on people of a particular time and place. People, by implication, spew out their expensive young the way a salmon spews out eggs, regardless of circumstance.

It is quite certain that all genetic strains of people who behaved in this prodigal way would have been suppressed by natural selection in the remotest antiquity. The expectation that it was physicians who engineered our population rise is, therefore, a direct denial that human habits have evolved through natural selection. If we were made by magic or special creation, or if the stork brings babies, then it is permissible to blame rapid population growth on physicians and sanitary engineers but not if we take a scientific view of the human species. The simple truth is still that the more spectacular part of our growth in numbers happened before health care, in the seventeenth and eighteenth centuries. Not only had the modern West reshaped the world before it began to use combustion machines, but it also had crowded Europe and peopled America with Europeans from a base of open sewers and physicians who believed in letting blood.

As western numbers and hopes rose together, so we went to war, but all our fighting made little difference to the map of Europe. We know the history of the last four hundred years in Europe to be one punctuated with violent wars and memorable battles. It includes the warring sagas of Louis XIV and Napoleon, of Gustavus Adolphus and Charles XII, of the Thirty Years War, of Peter the Great and Frederick the Great, of Bismarck, and of the two tremendous armies launched by Germany in this century. The European West has experienced the vastest and the most bloody wars of history. And yet the strange truth is that none of the aggressions which caused them has succeeded. Always, after initial triumphs, the aggressor has been thrown back. When the fighting was over the peoples of the warring countries were always left living in essentially their traditional lands. People were never displaced after aggressions as in the wars of antiquity. The map of Europe was never permanently changed by war.

Two things explain the failure of aggressions in the European West. One was that there have always been other ways of meeting the demands of the people than taking their neighbor's lands. Partly the demands have been met by the ever-increasing skills in agriculture and manufacture which let the swelling populations use the lands they already had to better advantage. But more important still was the presence of the New World to act as a people siphon, constantly draining off the excess people whose aspirations could not be met at home. The aggressive will of a people against its neighbor was damped in these ways, slackening the ruthless urge to rob which might sustain an aggressor king.

But the second reason for the failure of aggression is the stronger. It lay in the quality of Western weapons, and in the ability which literacy and communications

247

gave the peoples of Christendom to learn from one another. Any fit person could be trained to use a hand-
held firearm in a few weeks. Every European country
could copy and manufacture the latest cannon when
they learned about it. All the officers could read the
military textbooks of the other side, learning the lesson
of a defeat and preparing the logical counter. It was
thus never possible for an army of the European West
to maintain a decisive technological advantage.

The study of the great successful aggressions of antiquity shows that a decisive advance in military technique
and weaponry was vital to success in aggressive war. The
modern Europeans could never attain it. Each of their
cultures could master the techniques of contemporary
fighting, adapting them to their own particular needs.
Never again would it be possible for a way of life to
produce an invincible military instrument, as the steppe
nomad produced the disciplined horse-archer or the
Roman schooling produced a legion which could not be
duplicated in the tribal society of the surrounding barbarians. Any military edge which a clever European soldier could give his people would henceforth give them
the advantage for only a little while before their enemies
learned to emulate them or find a counter technique.
This fact is central to an understanding of the wars of
the last four hundred years.

Nevertheless, the attempts at aggression went on, in
spite of the repeated experience that it was hopeless and
in spite of the fact that the most pressing needs of the
people tended to be met by advances in technique or by
emigration to the new lands overseas.

We shall find that each of the most blatant of the
aggressions was from a wealthy and civilized state; indeed, from a state that was preeminent and proud in

civilized things. The France of Louis XIV, and again of Napoleon, was wealthy, admirable, cultivated. So was Spain when she had loosed her armies in the Low Countries and elsewhere, even as she took a large part of the American continent for her own. And the German armies of the last sixty years came out of a country dominant in art and science and in which the latest advances in industry were being made. All this is as the ecological theory predicts, for it is the wealthy and the well-off who are expected to undertake wars of aggression. Each of these countries is known to have had a rising population and rising aspirations. We also know that they were meeting many of the resulting needs for niche-space through technical ingenuity. Each of them exported people and in each the middle classes went most vigorously into trade. Each European aggressor, therefore, struck when wealth, numbers, and ambitions grew together, which is the condition for an attack predicted by the ecological hypothesis. Always the victim was a state smaller or poorer than the aggressor.

The same circumstance that drove these nations to aggressive war within Europe also let them try conquest in other continents. Each took lands from defenseless hunter-gatherers in America or Australia, from African nations equipped with iron but without firearms, and from Oriental countries so divorced from the learning of Christendom that they could not master the new ways of fighting quickly enough to avoid being overrun. But none of the aggressors managed to hold for long any of the land it took in Europe. This was true both for the deliberate wars of conquest started by established governments and for the wars of revolution in which the aggression was disguised and hidden as an exportable idea.

249

The European West has been well supplied with revolutionary wars. They came as the rapid winning of new resources made new ideas of liberty possible. When liberty flourishes, the old established ways will be challenged. The establishment always resists, and very often the young cannot make their idea of liberty the custom of the land without fighting. This means that revolution is likely to bring war, often even when it is only a revolution of the spirit with no immediate impact on the balance of political power. The Christian Reformation of gentle churchmen left Europe covered with corpses, particularly after its most frightful consequence, the Thirty Years' War, and the savagery of the French and Russian revolutions is something that every school child learns.

Wars in the name of revolution have often (though not always) been successful wars. The people knew what they were fighting for, which was to do as they pleased and to be governed as they pleased. They did not have to steal from others before they could claim the victory. They only had to persuade the soldiers of some of their own people that it was useless to oppose them further. At the end of such a victorious war the new order should be able to settle in the ancestral lands in peace. But it hardly ever does. Instead the revolutionary armies roll on outside the borders of their own state to attack their neighbors.

There is always a momentum to success in war. Those with the winning trick look outwards for another foe to strike down; and legion, phalanx or horse-archer fight on as far as their victories can be pressed. The rich and expanding society which forged the weaponry grants the soldiers their will, and finds political peace with the new opportunities won by the soldiers. A revolutionary war is a fight about a new niche inside the homeland, and victory in revolution has the same essential result as victory in conquest: resources are won for the new niche. More resources still can be won with yet more victory, and so the revolutionary army turns aggressor.

A revolutionary army, like those of Napoleonic France or Communist Russia, proclaims as its purpose the granting of new liberties to the lands it conquers. It sounds as if the army merely exports an idea. But the gains of its victories are material gains, and they are quickly seen to be so by its victims. The armies of the new freedom, reared out of relative progress, wealth, and expansion, turn into aggressor armies as the people promised fine things by the revolution seek those fine things in the plunder of other lands.

Each of the great continental powers tried aggression, either to expand the state or after revolting against the establishment. All failed in the end. The Hapsburgs, the Bourbons, the Republican Dutch, the Lutheran Swedes, the French of the tricolor, Germans in a united Reich, all tried, and all faced an inevitable hour when they were squashed back within their original frontiers. But the lesson was never learned; someone always tried again. The fates of these nations were set by the ecological breeding policy of the people, which forced their aspiring populations up, and by the realities of the modern technology of war in a literate polity, which said that no aggression could succeed.

The grand themes of the European West are thus recurrent revolutions brought about as liberty is made possible by gluts of resources, continual and ever-increasing growths of population, the siphoning of people to the New World which eased the drive to aggression, and recurrent tries at aggression all the same, none of which succeeded. These main themes can be followed more easily by tracing out the fortunes of individual nation states. The stories of Sweden, France, Germany, Japan and Britain show how all tries at conquest in the European West failed.

And yet the failure of European nations to beat down one another by force of arms, as the ancients were able to do, held only as long as the mutual knowledge of weapons prevented any power from achieving victory through a decisive and unbeatable technical trick. This essential condition holds no longer. With the invention of atomic weapons victory, of a sort, can be gained before the victim prepares a counter. Quite different scales of aggression are now possible, and the centuries of conduct according to the codes of the European West are now over.

Sweden

THE PENINSULA of Scandia has little farm land, much mountain and much winter, which means that it can support fewer people than flatter countries to the south. The Swedes who lived on much of it would be expected to feel the pressure early, if there was a general rise in number following the new inventions and the new agriculture. This seems to have been so. Swedes were among the first to drain marshes, to reclaim embayments from the sea, to use the new crop rotations, and to develop the system of committee government needed by the bureaucracies of a technically advanced state.

It does not come naturally to people of the later 1900s to think of Sweden as a turbulent, expansive, potentially imperial state, but that was the reality in 1600. The northern and eastern power in Europe in those days was not Russia, as it has been these last two centuries, but Sweden. Swedish trade was vigorous, and the wealth of her cities was growing, so that there were capital sums ready for new adventures. Swedish kings were able to pay hard cash for soldiers and weapons, although already under effective constraints of the merchant classes who had a real power of the purse over government.

Perhaps particularly telling of the strength of the middle classes was the enthusiasm with which Sweden embraced the social revolt of the Protestant Reformation. The new niche apparently suited the lives of contemporary Swedes as it did the merchant society of Tudor England, and Luther, who came out of a German church far away, was to press his mark more completely on Sweden than anywhere else. By 1600 Sweden had gone through social revolution, had revised ancient forms of government, and had collected capital. The people were conscious of the superiority of their material life over that of some of their neighbors and proud of the liberty of thought given by the new religion. It is not unreasonable to compare their country with ancient Athens poised to subdue her neighbors in Greece, the people trading, living well, and seeking still more for the younger generation. We know that both the population and the standard of living had risen in Sweden, suggesting that the country was ready for a war of conquest.

Sweden had for some time followed a policy of a limited show of military muscle and political leverage. This was beginning to yield extra land and resources from her neighbors round the Baltic Sea, when the series of fights which we know as the Thirty Years' War oozed out of Hapsburg Austria far to the south.

A Roman Catholic emperor had set upon Protestant-leaning neighbors and vassals with the self-interest of a robber baron and the rectitude of a missionary. He had as his instrument Wallenstein, a self-made millionaire, a genius of organization, a man whose shape we have seen shadowed in captains of industry of the recent past. Wallenstein procured the best army which money could buy, and proceeded to use it with the ruthless efficiency of his kind. The Austrian power scorched its way over the burning homes of reluctant people all across the plains of central Germany until it neared the Baltic, where it struck against the flexing purpose of the Swedish will.

The Austrian conquests were obnoxious to others besides Sweden. France, in the formidable person of Cardinal Richelieu, was ready to give money and material to any force that would check the Austrian power. Protestant England and others were equally interested in preventing an imperial Austria from debouching on to the North and Baltic Seas. The people of Sweden were ready for an aggressive war when, opportunely, the army of a distant state had come to ravage the land of her neighbors. Furthermore, this alien army came to impose a religious creed that could be painted as the enemy of liberty. When the Swedish armies set forth to do their national purpose their conquests would bring Sweden as many friends as enemies.

In all the history of the European West no moment was more favorable to the success of an aggressive war. And at this moment, Sweden and her armies were led by a warrior-king, Gustavus Adolphus, who has gone down in history as one of the greatest of captains. His soldiers set out into the heart of Germany, encountered those best armies which money could buy, first under the Austrian General Tilly, then under Wallenstein himself, and beat them utterly.

The armies which Wallenstein had purchased were modeled on those of Spain, the first country of the West to build a national army based on hand-held guns. They relied on the massed use of fire power. The soldiers were ranked and filed in big rectangular blocks, each with perhaps two thousand men in it. A block had its corners made of musketeers, square pegs of fire power that could cover the sides of the block, as the corner turrets protect a fortress, and the inner core of the block was made of pikemen who would stand against the rush of cavalry and hold the fortress while the musketeers reloaded. In this system two thousand men were made into a moving fortress, a fire-spitting phalanx that had been as terrible to the Moorish soldiers of southern Spain as had the phalanx of Macedon been to the Persians. The Spaniards called one of their moving fortresses "a battle," and it became the fashion for prestigious "battles" of their army to be commanded by princes of the royal house, the so-called "Infantas." The Spaniards had made the first of the typical military unit of the European West, the "battalion of infantry."

Swedish soldiers had worked out what to do about moving human fortresses; you made a checkerboard out of small blocks of musketeers and pikemen, the squares of which could cover each other at least as well as could the parts of a fortress block, and which also could be shuffled as the Romans had shuffled missile men with spearmen, and the Mongols had shuffled horse-archers with lancers. The Swedes, in short, met the organization of the phalanx with the organization of the legion.

But the Swedes also had an even more potent answer to the human fortress, one in keeping with the advanced technology of their aspiring country. They made light field cannon, which could be rushed across the battlefield with horses, and used them to smash the soft fleshy centers of the fortresses from a safe distance. The combination of legionary checkerboard and field artillery was to prove decisive, and at the battles of Breitenfeld and Lützen the massed battle blocks in the service of Austria were chopped away.

So decisive a technical advantage, such overwhelming victories, by the soldiers of a confident, literate, expanding state would, in other ages, have been sufficient to seal the success of an aggression. The Swedes had obtained the sort of triumph that Roman armies had obtained over Carthage. But they were not to enjoy the fruits of their victory as the Romans had done.

Neighboring peoples from all sides sent in their soldiers to quarrel over the spoils of Germany, and to help the cause of one side or the other, in what they took to be religious and enlightened self-interest. They rapidly copied the Swedish techniques and the proliferating cannon and muskets shot each other's soldiers away, until the people were sick of it. Northern Germany was made into a largely depopulated waste. Then, at last, a peace conference was called. Powerful neighbors each held onto border crusts of the German loaf, and the rest was handed back to the impoverished German survivors. Christendom had been restored to its usual state of a balance of subsidiary powers, and a lesson about the hopelessness of attempting annihilating conquests within Christendom had been learned which was to guide Europeans for nearly a century. The memory of the Thirty Years' War was to impose a code of conduct on European armies that restrained the fury of war until the creeping exactions of French armies of Louis XIV provoked Marlborough's terrible answers.

But Sweden's military policy had not been entirely in vain, because in the general peace conference she had been allowed to retain coastal provinces of North Germany, which gave her an outlet, both as a place to administer and, through harbors on the North Sea, to world trade. And Sweden, as a victor and a preeminent military state, could press on with her policy of expansion through limited local success. Particularly could she do so toward the east, for in that direction were the vast lands of the Muscovites, scarcely parts of Christendom, technologically backward, apparently no more capable of finding an answer to musketry legions than had barbarian kings been able to find an answer to the javelin-throwing prototype.

For many years Sweden was to follow a policy of annexing land to the south and east, taking over small states, interfering with her efficient armies in the squabbles of Muscovites and Poles, steadily winning territory and consolidating a northern state which looked like remaining one of the greatest powers. Sweden came to be, not the long narrow country we think of, but a spreading disk of land, with a hole in its center which was the Baltic Sea.

Swedish life changed as the empire grew, telling the tale of expanding niches and numbers. Trade flourished. Painters and writers experimented, giving clear sign that many people lived in broad niches of their choice. As the *Cambridge Modern History* puts it, "The nation seemed to be struggling to fit itself for the great position which it owed to the fortunes of war and politics." Niche theory suggests that it was individuals who were struggling for a better life, not the nation for its mission, but the historical facts of a boom in life style seem clear enough.

Yet Sweden also came under increasing despotism as the king, head of the powerful armies, got full control over the privy purse. Repression of people by bureaucrats and religious police tightened, but these could be endured for a while since Sweden got steadily richer, in spite of the costs of king and Church. Control of the Baltic Sea gave Sweden a monopoly of trade routes from Asia to the West, and she reinforced the monopoly by strong protective tariffs. Swedish people must have felt that their victory was truly won and the new empire secure, yet the forces which made empire in Europe impossible were even then working on the eastern frontier and were soon to sweep Swedish triumph away.

Russia came under the dynasty of the Romanoffs, kings rough enough in their manners to understand their half-civilized subjects, but schooled enough to know that their people must learn the techniques of the European West. They began to import technology and skills, and the fourth of them, Peter the Great, went about this task with the tenacity of genius. It was well for the Russians that he did, because his efforts to find a trading outlet on the Baltic brought on general war. Soon Polish soldiers had bitten their way to Moscow and another brilliant Swedish soldier-king, Charles XII, was cutting up the Poles on his way to making the Russian lands part of Sweden.

Weapons and military techniques had changed a little since Sweden had begun to build her empire, though none of the new ideas was as decisive as those with which Gustavus Adolphus built it. Bayonets had been invented, letting each soldier be both musketeer and pikeman, yet the Swedish army still used pikemen in something like the old array; an eighteenth-century illustration shows Swedish formations bristling with pikes as they advanced on Russian infantry. Sweden, perhaps, may not have had quite the latest thing in weaponry, something not unknown in armies of other long-victorious empires. But the Swedes did have the massed cannon and muskets in legionary array, which they themselves had pioneered, and they had the martial organization of a professional army. These were ample to secure their empire in the East unless some force of comparable technique could be brought against it.

In Charles XII, Sweden had a soldier's soldier, a dashing, inspired, dauntless man, with an instinct for tactical advantage almost magical. He was everybody's dream of a hero-prince, a Roland, an Oliver, a Richard the Lion-Hearted, and, perhaps more ominous for Sweden's future, a General Custer. His citizen soldiers followed him from triumph to triumph across the Russian lands.

But Peter Romanoff studied Charles, schooled himself in Swedish tactics, imported muskets and cannon, threw aside the privileges of rank in his army and made officers of the rough friends who understood him. He lured the dauntless Charles far into Russian territory, where supply lines were stretched and General Winter could get a grip. Then he stood; with an army four times as big as that of the Swedes, armed with the weapons of the European West, arranged and dug in as the best Western writers said it should be. The Russian king waited for his attacker thus at the little town of Poltava, in the Ukraine, part of his army in the town, part a short distance off. Charles XII needed Poltava. The winter was very hard and his soldiers had fought their way all across the neck of Europe, an appalling distance. They were cold, hungry, and far from home. Inside Poltava was shelter and food, but the defenders outnumbered the besiegers and the town fortress held. Then Peter brought his main army, with weapons as good as the Swedish weapons, and pinched the Swedes between two forts. He built a rectangular redoubt a little way from Poltava town and waited inside it. There was a river on one side of this fort, and the road to the front of it from Poltava was pressed between two blocks of woodland. So Peter started building little forts with cannon to command the pass between the woods; lots of cannon; imported cannon; more cannon than were left to the whole Swedish expedition.

The cold and hungry Swedes were now between an army in a fort and an army in a town. But they were used to smashing the barbaric Russians when the odds against them were worse than five to one, and they followed the dauntless Charles with spirit as he determined to rush down on the Muscovite mass in the fort; their civilized ways should carry all before them as in the past. This was the spirit in which Custer rushed his cavalry into the dust cloud on the Little Big-Horn. The result was the same.

The Swedes, in disciplined lines of foot and horse, swept into the passage through the woods, in two files, on either side of the little forts built there. One line got through, but the other was torn and split by rolling rods of cannon balls until their formation was gone and a Russian sally surrounded the huddled remnants and compelled surrender. The second column cleared the cannon, hesitated, got their discipline back, and stared at the main earthworks behind which the Russian masses waited. Then the memory of so much victory took them and they charged. But it was again into Western cannon that they rushed; long rows of cannon hidden behind the earthworks, perhaps not fired in pretty salvos, but manned by people who knew enough to ram in powder and light a match. They fired canisters of bullets, grape, and small shot. The Swedish army disappeared.

The Battle of Poltava was one of the most decisive military events of all time. It meant, of course, that there would be no Swedish empire and that Russia would hold her outlet on the Baltic; but this, though very important, was still minor. The real decisive message of Poltava was that, in a literate world possessing firearms, lasting aggressive conquest was impossible. The other side, or the other side's friends, or the other side's neighbors, could always get firearms for themselves and turn the temporary conquerors back again.

France

FRANCE, IN the second half of the seventeenth century, was the richest, the most populous, the most technically advanced nation in Europe. Her industrious people prospered and multiplied, and French aspirations were high, as they could afford to be. Not only could broad, fertile France find room for a large number of organized and ingenious people, but wise governments began to win for France a share in the Americas, in tropical islands, and in the subjection of India. The wealthier French classes took to trade and the governing of colonies. A powerful central government gave employment in the civil service. A large standing army was needed to maintain the apparatus of the state, giving living to both the masses who filled its ranks and those who supplied them with arms and supplies.

The ecological hypothesis suggests that this France ought to have rising numbers and increasing hopes of broad niches for much of the population. The record leaves no doubt that both were happening, for we have data on population, on the growth of cities, and on the broadening of the ways of life to show that these things were so. Class repression should be very visible, and again the record shows that it was. And the hypothesis also suggests that such a country would soon engage in aggressive war against its neighbors. Under Louis XIV the aggression came.

French rulers began to squeeze the governments of France's little neighbors, using threats, money, and subversion, in ways not unfamiliar to enlightened peoples of the twentieth century. It was, of course, necessary to back up this pressure with military force. France used her money to maintain the largest standing army in Europe, and her ingenuity to make it the best-equipped and led by the most learned officers. Bayonets let the French discard pikes and give all their men firearms; they made better and better cannon; they secured political gains with marvelous fortresses, over which their most gifted engineers and architects seem to have had a most enviable free hand. One by one the smaller powers were to be expected to submit to this glittering force and the French writ would thrust its way across Europe. The reward to the people would be "glory"—a nebulous dimension of niche like "liberty" but equally important to people who have learned to want it. There can be little doubt that the push to conquest was popular with most of the French.

But, in spite of their military preparations, the French were anxious to avoid total war. All Europe kept fresh the memories of the Thirty Years' War, its horrible slaughter and its shocking withdrawals from the treasury. France was not to be let in for anything like that. So French soldierly writers put forward a theory of limited war for which their beautiful military toys were to be used. There should be no unnecessary savagery about their fighting, for they were civilized and would not want unnecessary killing or damage to property.

The object of "limited war" was "the settling of disputes." When you fought against your little neighbor, you were not engaged in aggression, you did not go to rob him. No, you were simply doing your duty and helping to "settle a dispute." We must not dismiss this fatuous thinking as merely an aberration of the eighteenth-century French, for you will find it echoed in the apologies of modern soldiers too. It is still taught in the war academies. It is the sort of folly which was in Clemenceau's mind when he said that war was too serious a business to be left to the soldiers.

The disputes which France proceeded to settle by "limited war" were over whether, or not, there should be more goodies for France. French wars tended to have causes like deciding which royal house had the right to a throne, hence names like "The War of the

Spanish Succession." What was really in dispute, however, was always the French right to some of the property or power involved in the "succession." France appealed these disputes to her judge of battle, and since she had the most money to pay the judge, the disputes at first tended to be settled in her favor. But there was another way to appeal this judge's decisions than giving her more money; you could stop limiting the war and force an issue through stricken armies and dead soldiers. The victims of France were, at the last, driven to this harsh form of appeal.

The rock on which the eighteenth-century French wave of aggression broke was hewn out of an alliance between Holland and Britain, at first led by William III, Prince of Orange, and then by Marlborough under Queen Anne. Britain and Holland were trading nations, populous, full of merchants, rich with a capital collected by trade though poor in resources within their homelands, having turned their comparatively small patches of land into intensively farmed gardens and holding the sea lanes with powerful war fleets as parts of their national traditions.

As the appetite and military power of France grew, so King William and Marlborough saw themselves maneuvered into the fatal position of a Carthage before a greedy Roman republic. Once let the French power hold the rest of Europe, then it would be only a matter of time before the resources of the continent were consolidated against them. Larger, more powerful fleets than their own would hold the sea lanes, and the commerce which maintained their capital and their wealth would be taken from them. Their peoples might not be annihilated as were the Carthaginian people in a similar plight, for such was not the custom of Christendom, but they would be impoverished and their countries would become the exploited provinces of a French superstate. In Marlborough's letters home we find this dismal consequence of failure clearly spelled out.

Marlborough was given command of both Dutch and British armies with orders to stop the French. As he brought his small armies against the French, his problem became Hannibal's problem. His policy must be to make the French let go even as Hannibal had tried in vain to make the Romans let go. Yet Marlborough was in nothing like Hannibal's plight. Britain and Holland were not all alone to face a giant power among crowds of impassive or hostile barbarian tribes. On the contrary, they were to struggle in the heart of Christendom, and with civilized and by no means helpless powers on every flank of France. So the first step in resistance must be to secure friends, to get help against France. Where Carthage had had to face a continental fury all alone, the British and the Dutch sheltered in the shifting patterns of the Grand Alliance.

The second thing in Marlborough's favor was that he could match the military technique of the enemy. Not for him the pending shadow of a Zama where brute superiority of military technique could clinically set aside the advantages of valor and generalship. The British and the Dutch could learn and imitate every technical improvement of the French. All knowledge was common knowledge. Every regiment that Marlborough raised could be confident that its weapons and its drill were at least as good as anything which could be brought against it. As it happened they were better. The French army, perhaps through the inertia to which centralized organizations are prone, was slow to replace their matchlock muskets. The British and Dutch issued flintlocks first. They could fire more volleys in the same time. This is the sort of thing which inspires confidence in soldiers. In Marlborough, the soldiers had the sort of general who could take advantage of both those weapons and that confidence. He knew it. He worked patiently for his chance, then forced it to come, far away in the heart of Europe, near the Danube at the little village of Blenheim.

By 1704, the year of the Danube adventure, France had been pressing her aggressions to the west for more than a quarter of a century, and the nibbles of land stuck on to France were growing into bites. French armies were in what is now Belgium and looming over the Dutch Republic. To the south, other French armies lapped at the river Rhine along its length, and crossed it. Then Bavaria declared for France, as its ruler decided to opt for safety in appeasement. French armies were now in the heart of Germanic Central Europe. It needed only some other smaller German states to follow Bavaria in collaboration with the aggressor for a French empire almost to be made.

And caution in the fighting had still been the guiding spirits of all the generals. French skill at maneuver, coupled with the frightening size of her armies, had given the successes. Her soldiers had grown used to winning without desperate loss, and the armies of France were filled with soldiers who were loyal, even after all these years of war, because their generals had been sparing of human life. In following the dictates of limited war, French commanders had automatically followed that precious prime dictum of good generalship, not to squander the lives of your men. The French armies were in very good heart.

Governments in the Grand Alliance were equally cautious, particularly the Dutch who had the most to lose. For them a knock-down, drawn-out, killing battle might be the last disaster. They were small, and France was big. And the conditions of fighting in the musketry wars were bound to turn any efforts at decisive victory into an affair of horrible losses. A decisive fight would perhaps be worse than battles of the Thirty Years War, because now every foot soldier had a musket, and the opposing ranks had to shoot each other down as they stood, face to face, thirty or forty yards apart.

Yet General Marlborough and a very tough-minded Austrian politician, John Wratislaw, resolved that the thing had to be done. They chose the French army in Bavaria as their target. A fight in Bavaria would be over the land of the one German prince who had turned appeaser, threatening to start a rot in the Grand Alliance. Let his country see, as a warning to others, how dreadful war could be. But, more important still, this remote battle zone of Bavaria would leave the battle to the generals, out of reach of politicians in London and The Hague who might call them off at the last minute, as they had done before.

Marlborough's British army walked from Holland, three hundred miles through friendly German lands, across the Rhine at Coblenz, past Mainz and Heidelberg, across the northern reaches of the Black Forest, into the appeaser's land of Bavaria, and along the Danube from Ulm to Donauwörth. There the British joined an allied German army from Baden and stormed the Schellenberg fortress which guarded the town. They stormed it the day they arrived, marching grimly at its ramparts, the British toward one side, the Baden Germans toward the other, having a third of their men and most of the attacking officers killed rather than fail to secure a fortress needed as their base, so far from home were they.

Then the British and Baden Germans began setting fire to all the crops and villages of Bavaria. A burning land; this was to be the price of appeasement. And the burning would surely draw out the French regiments to where they could be fought. The British were joined by an Austrian army under general Eugene, an elected prince of the Holy Roman Empire and a man sharing Marlborough's hard vision of what must be done. Together they came upon a great French army near the village of Blenheim.

Now was to be fought one of those frightful battles where the two sides have the same weapons, the same training, the same tactics, and the same courage. The French were brilliant soldiers from Europe's premier state, confident, unbeaten in a quarter of a century of war. But they had no unbeatable technique as real conquerors do. They had come to this high tide of French imperial power not through crushing their victims but by outsmarting them. Now this tactical bluff was about to be called. Eugene, Marlborough, and all their officers were bent on destruction and death.

The French line ran from Blenheim village on the bank of the Danube River inland to broken, hilly, wooded ground, a distance of about four miles, a very secure and defensible position for an army of fifty thousand, well supplied with cannon. Running straight across the front was a stream which flowed into the Danube; in front of the middle of the French line was a marsh, a most serious obstacle for infantry who must keep formation as they move or be lost. On the right, therefore, the French flank rested on the Danube, secure, unturnable, and given the added strength of a fortified village. On the left their flank rested on wooded hills, and was equally secure because eighteenth-century armies could not march through woods, nor could cannon shoot through them. And to the front, the French could be attacked only across the obstacle of a stream and a marsh; a very strong position for a powerful army set to fight a defensive battle. But the French generals had one great weakness as they planned the defense; they did not know the savagery and purpose which was swelling against them. It even seems that their generals did not really believe they would be attacked at all. They all slept in their tents, keeping a thousand yards back from the little stream in its unpleasant marsh. Had they fortified the edge of the marsh, the day might have ended differently. But Eugene and Marlborough deployed their armies early, in the dark, and, when the sun was fully up, British redcoats and Austrian gray were spreading left and right along the allies' side of the stream, from the river to the woods. The British could still be pinned to the marsh with cannon fire but it was now too late for the French to defend it closely.

Then, on the flank by the River Danube, the British soldiers walked forward by companies, in parallel array, as if on parade, muskets shouldered, under orders to fire no shot until within thirty yards of the palisade round Blenheim village. Cannon balls tore lanes down the marching companies, spaces emptied of living men, but they closed ranks and walked on. Many hundreds of French muskets fired at once through loopholes in their palisade and one third of the British were dead. The remainder rushed the village, trying to fire back through the loopholes. Every British officer was killed as he struggled to climb the palisade. But behind came more British companies and more; they were shot down; it was unprecedented butchery; but a redcoated sea still lapped round the village and there were mounds of French dead inside to offset the British mounds without. A French officer seems to have lost his nerve then because he called in reinforcements; and the message got back to Marlborough that large parts of the French army were pushing to the defense of Blenheim village. His ruthless plan was working, the French were taking soldiers from their center to counter the storm at their flank. To the survivors of his regiments from the bloody assault at the palisades Marlborough's only message was to keep attacking, to hold the French reinforcements in the village.

Meanwhile Prince Eugene's Austrians stamped with equal fury into the other end of the French line. They had the woods and hills on their flank; rough country, fortified, palisaded. Here it was gray companies that marched on with the same deadly intent, struck down by the trained volleys of long French lines, answering with similar volleys, line against line only thirty yards apart in dense powder smoke, with the eternal flashes and the soft lead bullets that mushroomed as they struck. The French threw counterattacks on Eugene's flank too, battalion after reserve battalion, but still the Austrian lines came on. It was a new frightfulness. The brave Eugene sent his messages of confidence back to Marlborough in the center, "All my soldiers are committed, but tell me if you yet need help." But Marlborough replied, "Keep pressing, hold them there."

It is the impersonal will of an army that sends soldiers to walk to death in scenes as horrible as these. Picture yourself in the Austrian or British ranks, tramping in your company toward the crowding French strongholds at either end of Blenheim field. You walk, as you have often walked before, in company that you know, perhaps toward death. But not everyone will be killed. The better your fire drill, the fewer of you will die. You are, anyway, *better* than the enemy. At least the cavalry won't get you. While you are in your ranks and have your bayonette no horseflesh can be forced onto you. If you lose your company through panic in front of the French though, their cavalry will kill you for sure. A running soldier is a dead soldier on a field where there are horsemen about; sabering runaways is what cavalry are for. If you broke ranks *before* you got to the French your own cavalry would kill you. Your best chance is to keep closed up; and fire drill; drill, think of your drill.

So the red and gray waves walked, and drilled, and shot, and were smashed by the heavy lead balls; but they held the French soldiers to the redoubts on the wings. And in the center the British infantry was bridging the stream and wading the marsh; forming up the other side to stand, pounded by cannon, hour after hour, until they attained that local superiority of three to one which is decisive. Then forward, and split the French line and the French army in two. And the Austrian and British cavalry tore on through the gap. The French masses in the wings, at Blenheim and in the woods, were surrounded and taken, those of them who were still alive. The First Army of France was gone in a single day.

Marlborough did it again, in other battles, in the following years, at Ramillies, Oudenaarde, and Malplaquet, though it was harder and harder each time, for the French were learning and were fighting, at the last, in France itself. No battle gave final victory, but the beating that Frenchmen took was enough. The enemies of France all took heart, and the French king, much of his treasure spent, looked out to see on all his frontiers the angry peoples of Christendom with weapons in their hands. The French let go, even as the brave Hannibal had hoped in vain that the Romans might let go two thousand years before.

This experiment with aggression by limited war had lasting effects on France. The French used up their savings in the enterprise, spending capital to buy armies which would have been better spent in making provision for the rising aspirations of a rising number of people. Worse, she lost ground in the race of Western powers to secure the new territories overseas. The kinds of aggression which *did* work for the European West were those against stone-age American Indians or declining Asiatic nations who could not match the modern European armies. These aggressions won the land which gave the Europeans a siphon for their bounding populations, or resources to provide the new life for those who stayed at home. By her vain try for European conquest, France let go some of the plunder of the outside world to the Dutch and the British.

Then France was left with the largest and most aspiring population in Europe, penned within her national boundaries, with dampened chances for expansion overseas, frustrated in war and balked of the plenty their leaders had promised them. Now the French leaders looked to themselves. The aristocrats and the churchmen sought refuge in privilege, and the rising numbers of would-be middle-class people were made to pay the taxes and alone endure the changes needed to accommodate French society to its new condition of rising numbers, rising wealth in material things, but a fixed supply of land. The French caste system took on that aristocratic arrogance which is still legendary.

But this was still a time of great changes in the condition of life, changes brought about by the release of wealth through technical ingenuity. It was not the stage late in the history of a civilization, when a caste system can be endured because seen to be necessary. It was, on the contrary, a time of rising hope, when the new plenties could be sensed on all sides. It was a time when the young would talk of their new things and call them "liberty."

Even the law-and-order middle classes found themselves critical of the social repression of those in charge of the state. They were able to turn their resentment of high taxes and irksome changes of habit onto the privileged few who maintained old customs at everyone's expense. The taxes and the changes were an essential part of organized bureaucratic life required in a nation of high technology. They were made necessary by rising numbers and economic growth. But the French system of rank gave the middle classes an obvious scapegoat. What was happening was so clear that even French authority knew it, and a French king, Louis XV said, "After me the Deluge." In his successor's reign the deluge came.

There was no money to pay the interest on the national debt; the French Parliament was called after a long vacation of 160 years; the king moved up some regiments to be ready for trouble; the people reacted to this insulting show of force in the way which might have been expected of them; the bastille was stormed; the king fled and was brought back a prisoner of the Parliament. Then his friends on other thrones saw a chance of doing themselves a bit of good by helping him back to his. A joint army of Austrians and Prussians accordingly invaded France. It was a rather fine army, nicely drilled, beautifully equipped, knowing all the elegant

tactics of limited war. Its commanding generals were
famed for their skill in maneuver, one of them boasting
of having waged war for a whole campaigning season
without ever fighting a battle. With the beauty of their
drill they were going to sweep away the rabble (their
spokesman used the word "rabble" repeatedly); then
they were going to put Louis XVI back on his throne
and appropriate some of his outer provinces for "services rendered."

The pretty Austro-Prussian army met a gathering of
French soldiers at a place called Valmy, where the
French stood on a low hill, sixteen thousand strong and
with forty cannon. There were more in the neighborhood, but it was the ones on the hill that the invading
soldiers decided to push away as rabble. In embarking
on this enterprise the invading generals showed that
curious lack of understanding of why people fight inherent in the more pedestrian sort of soldier. It is true
that some French formations had run away in the
months preceding. The French had had to put up with
being used as cannon fodder by arrogant officers in the
past, and were not ready to stand against bullets unless
they thought it worthwhile. But at Valmy it was plain to
them that fighting was worthwhile, for running from
the Valmy hill meant a return to the old caste system.
Their new general, Dumouriez, had put it to them
clearly enough. "Do you think that liberty can be won
without fighting?" he said, a little figure in an open
field, shouting before ranks upon ranks of doubtful soldiers. For the French, "liberty" meant the new niche,
the new way of life, just as it had meant for the framers
of the American Constitution. The Frenchmen on
Valmy Hill had decided to fight; and they held first-
class muskets and cannon with which to put their will

into effect. Possibly their buttons weren't polished, and this led the Austro-Prussian intelligence people astray.

The battle opened with an exchange of cannon fire at long range between the commanding height known as "La Lune," which was held by the Prussians, and the hill of Valmy. It is said that twenty thousand rounds were fired on each side, the most intensive bombardment yet, and which has let historians refer to the affair as the "cannonade of Valmy." The French stood. Indeed, they probably got the better of it, because the French guns were the best in Europe—a fact which all their neighbors were soon to learn. Then the pretty parade-ground lines of Austrians and Prussians, their brightly polished buttons doubtless gleaming in the sun, were set to march up the hill and push the rabble off the top. But the rabble didn't go. Historians tell us they shouted, "Vive la Nation," then fired their guns.

The invaders never reached the top of the hill. They turned back and made their way home to Austria and Prussia. The poet Goethe had gone with the Prussian army to see what war was like, and that evening he dined in a Prussian officers' mess. He wrote that the officers were very silent and, at last, when he was asked for his opinion of what had happened, he replied, "A new era in the world's history has just started, and you can all say that you were present at its birth."

For the French, the new era found a citizen army fighting an aggressive war for resources to live better. The people had seen the chance of a better way of life for their children. This chance had come from technical changes, from the new agriculture and the new industry, and from the immense resources which growing communications let them tap through trade. But the promised good times were slow in coming. The people had removed one obstacle, the caste system of their governors, and distributed the wealth of aristocrats and churches among themselves, only to find that there really was not as much to loot as they had been led to believe. At Valmy they learned to fight for the little that they had already. It was but a short step to learning to fight for more.

The mass was ready to battle for loot; soon administrators and officers were ready to battle for new territory. A message from Paris was shortly to go to the French commander in Italy:

The Executive Directory is sure, Citizen General, that you consider the glory of the arts to be linked with that of the armies you command. . . . Italy, in large measure, owes to them her wealth and fame, but the time has arrived when their sway should pass to France to strengthen and embellish the rule of liberty. Our National Museum ought to contain the most celebrated examples of all the arts, and you will not neglect to enrich it with all that it may expect from the conquests of the army of Italy and from those of the future.

It could not be long before a people so resolved would throw up a great captain to lead them, as had others in similar circumstances. The name, this time, happened to be Napoleon.

Republican France started with two military advantages: the most accurate and mobile field guns of any army, and soldiers individually resolved to conquer. There were no other advantages, no technical advance which should make her armies necessarily victorious when properly handled, as the Macedonians, the Romans, or even the Swedes of Gustavus Adolphus, were. Yet the resolve on a total war to conquer, channelled by

276

a soldier of genius against armies of states less certain about their purpose, was to produce so much victory that a dynamic of aggression rolled out from France, as it had from ancient conquering military republics.

Napoleon organized his foot soldiers as battalions of musketeers of about ten ranks with seventy men to a rank. Each battalion was an oblong which marched into battle sideways with seventy men facing the enemy. Napoleon stacked numerous battalions, one behind the other, each with its front of seventy men, so that there was a long column of rectangular blocks, like the shaft of a giant's spear seventy men thick. When Napoleon thrust, this spear rolled forward shouting, "Vive l'Empereur." But first he blasted a hole for it with the inspired use of cannon, cavalry and skirmishing riflemen; and he ensured for his spear that numerical superiority at the decisive point of two or three to one which the general of genius contrives for his men. Italy, Austria, Prussia and numerous lesser states fell before Napoleon's ardent spear; the names of Austerlitz, Jena and Auerstadt are ones that soldiers will ponder until war is forgotten.

But all was in vain. The victories glitter down the years, but the conquests withered away. France had no technical superiority. The initial advantage of having soldiers determined to conquer quickly went as her annexations instilled an equal desire to resist into her victims. The French armies moved across a Europe which was literate and technically advanced, where people could study the French success and prepare the weapons and the tactics of a repulse. Britain resisted as she had resisted the earlier French aggressions, and for the same reasons. German nations armed and rearmed. And the Russian power could not be struck down at all.

The British found the tactical answer to the attack in column; you spread out highly trained musketeers, and even riflemen, on a long front enveloping the spearhead, then shoot it away. The Russians summoned General Winter and prudent, patient generalship to their aid and were able in 1812 to waste away the whole French army. Three years later France was back to the size and shape it had always been.

Germany

ONE MAN, above all others, understood the meaning of the thrust and parry of the aggressions from revolutionary France: a German soldier, Karl von Clausewitz. Clausewitz saw the whole thing through, first as a young Prussian officer in the armies shattered by French resolve and Napoleon's genius; then as a staff officer with Scharnhorst, when they planned the military instrument which a now angered German people were to use against Napoleon, and at last on secondment with the Russian armies during the ruthless, total and annihilating campaign of 1812.

Clausewitz had seen that the idea of any limit to war became absurd when the soldiers were sent out by people bent on real aggression. He felt, through and through, that advanced weapons were now things anybody could make and use. Wars would be fought with common weapons and common tactics, so it followed that only strength of resolve and weight of numbers would decide the issue. Wise policy could only be to foresee war, to prepare for it, and to try to provide good generalship when it came. Clausewitz's textbook *On War* was published in 1832, in the decade when Germans set about putting together their modern nation-state.

When the Napoleonic wars were over, German people were left scattered in thirty-eight, mostly tiny, German-speaking, neighboring states. The people were the industrious Germans we know, already turning their splendid craftsmanship toward the new industries. In a peaceful Europe both their numbers and their aspirations grew. But the new industrial ways required large markets and this a thirty-eighth part of Germany could not provide. The people's aspirations were frustrated, and very large numbers of them took advantage of the American people-siphon and quit German states that were too crowded for the contemporary niche. But others saw that the chance of a better life through industry would increase with a common market. It was particularly important for them to be able to protect the labors of German craftsmen against the products of British mass industry, which began to dominate German markets when peace came. The first, and obvious, response was Customs Union, a German common market. Politicians are slow about these things, and it took them thirty years to engineer a Customs Union of all the Germans, but by 1845 they had done it.

Prussia was always the biggest German state (outside the Austrian Empire). It was in Prussia that Clausewitz wrote, and for Prussia that Scharnhorst had worked to model a new army for the defeat of Napoleon. Scharnhorst and Clausewitz had followed the logic that in a war which only numbers and resolve counted, the governors must do their part by preparation. Multitudes of soldiers, armed with modern weapons, needed an immense administrative bureaucracy; they needed rule by committee. Scharnhorst had provided the first committee in the form of the German general staff, and the readers of Clausewitz perfected it. The wisest of Clausewitz's students, von Moltke, found himself Chief of this General Staff, when a Prussian king called Wilhelm and a prime minister called Bismarck addressed themselves to the task of turning the Customs Union into political union.

This was a task understandably popular with many Germans. It required no war of conquest and, indeed, no war at all were it not for the fact that the other European powers liked Germany fragmented. But the Germans had an army, inherited from Prussia and rapidly swollen; large, well-armed, governed by a skillful committee, and made up of soldiers whose way of life was likely to be improved if German union was preserved. The clever Bismarck tricked his neighbors into attacking him, one at a time for safety, first Austria and then France. Bismarck used ancient grievances about border provinces as baits for his diplomatic traps, Schleswig-Holstein and Venice to infuriate Austria; Alsace and Lorraine to madden France. These were leftover territories after the general settlements of 1815, lands of ancient dispute whose loss would always be a grievance to some power, somewhere, sources of brooding discontent not only to rulers but to common people also. Bismarck worked like a knave among hotheads, teasing one about honor, praising another about justice, until he made his two victims angry enough to attack his new and shaky confederation of little German states. In this way Bismarck gave his Germans a common enemy and a chance to see how superior their army was, both at the same time.

The attacking armies of Austria and France were as large as those von Moltke wielded for Bismarck, and they had weapons nearly as good, but their soldiers were sent on what amounted to a half-baked aggression for which their resolve could not have been very great. More importantly, they were without the organizational backing of efficient general staffs. The German soldiers, confident, well fed, well equipped, well looked-after by their meticulous committee, defeated Austria at Sadowa and France at Sedan, and went home again. Victory went to German heads a bit, and in the 1870 war they grabbed a few provinces like Alsace and Lorraine as loot, but they hadn't really been engaged in aggression. What they had done was to put together a Germany big enough to give them all room in which to live well for a time.

The German union was now complete, and an industrious people proceeded to show what they could do with the new techniques, striving to raise the aspirations of all. But the numbers grew also, and the German lot became a fatal race between babies born and new resources won. The Germans took full advantage of the American people-siphon, and peopled whole states of the American Union during these years. They also turned to trade, seeking to buy the resources of distant lands with their manufactures. This was also the resort of every advanced country in Europe. The resources of trade must be won against competition, and here Germany's fragmented past had left her people at a disadvantage. The most desirable outlets for trade were claimed by others as wholly owned subsidiary properties. Britain, France, the Netherlands and the rest looked upon themselves as "holding companies" possessing a very large proportion of the whole earth as trading assets. They seemed able to provide for many of the needs of their expanding populations with the resources of these assets, and thinking Germans, charged with their country's affairs, were persuaded that the German standard of life could be maintained only if Germany did the same. It was to be German policy to be ready to secure a share of overseas territory when opportunity came.

Germany's own Clausewitz had said that war was "policy by other means," and this was true as long as your policy was to be to take what others had. Thinking Germans could see the needs of their country growing, could see the numbers rising, and the per capita demand rising too. They gave no thought to changing the breeding strategy, and simple logic told them that the day must come when they would need more trade outlets, more home territory, or both. As Clausewitz had taught them, they maintained their national army and the excellent ruling committees of a general staff that so large a force must have. They didn't think of themselves as bent on aggression, only of being prepared to take a fair share of what the others had parceled out during Germany's years of weakness. Even the need for this might not come soon; it might be many, many years before the wants of the people said, "Get it for us now." But it was just as well to be ready; so Germany got ready.

The above I believe to be a fair statement of what a newspaper-reading German of the period might have said of German policy. The Kaiser made speeches about military virtue, merchant politicians and editorial writers noted the British hegemony of trade in her empire, the German Foreign Office struggled for colonial space in Africa, and universal conscription made all Germans take their army seriously. On a niche-theory view of history, the public was clearly right to expect war. Expanding wealth, expanding numbers, individual hope, a superstate recently put together, trade, a tightening class structure—all the symptoms were there. What was lacking was a suitable victim.

Germany's continental neighbors were France, the Austro-Hungarian Empire and Russia, each a large continental power, each with universal conscription, each armed as the Germans were armed. All three were keenly aware of what had happened in Bismarck's wars, France and Austria having been Bismarck's victims. They built their own general staffs on the German pattern; their officers were trained on Clausewitz; they read the wise von Moltke's own account of how he had destroyed the France of 1870.

France and Austria, far from being set as classic victims of an aggression, had the rising wealth and aspirations which let them think of conquest themselves, France in Africa, Austria in the Balkans. Russia's lands and backward technology might in some ways fit the pattern of a victim, but her army was the largest of all. So the heavily armed Europe that faced the start of the twentieth century was made up of large powers each of which was wealthy, thriving, possessed of advanced technology in war; and with both broadening niches and bounding populations at the same time. Each, therefore, was in the condition described by the ecological hypothesis as necessary for aggressive war. Germany was merely marginally the most powerful, and the state with the most grievance against the others.

With aggressions all set and ready to go, war came by accident. A worthless little assassin at Sarajevo shot a rather decent Archduke and his even more decent wife. One power accused another of paying the assassin, people in minority communities exulted at the murder, as their modern counterparts with acronyms (PLO, IRA, ETA) do at similar foul deeds, and general staffs used the ponderous communications of the day to mobilize their soldiers. No actual deliberate aggression started the First World War, even though potential aggressors were plentiful enough. Dull statesmen fumbled with national armies, got them fighting, and found that they had populations preconditioned for war who would not then easily let go.

The war was fought with machine guns, and artillery that fired explosive shells. All combatants fought with similar weapons. No side had a decisive advantage in technique. The real novelty of the war was the organization of general staffs, the bureaucratic committee approach to fighting that Clausewitz had pioneered. This meant that people could be maintained on the battlefield for very long times, rather as similar bureaucracies maintain people crowded in cities. A battle could now last for weeks or months, instead of the old norm of a day or so. But since all the powers had similar general staffs, this made no difference to the outcome.

In desperation to find a way of surviving through machine-gun fire, the tank was thought of in all the armies, the Russian, the German, the French and, most particularly, the British. Lone British soldiers in muddy trenches were dreaming of how tanks and aircraft might be put together to achieve a whole new concept of war. But that was for the future. This fight went on with mutual killings until revulsion began to take hold in each populace.

The Russian civilization broke first. This was to be expected because Russia had less per capita wealth than the other powers. Russia already had a great land empire and her industrial base lagged several decades behind that of Germany. Too many of her people were poor for there to be much appetite for a war of expansion, and even the bureaucracy and the general staff were inefficient by the standards of the other nations. So, alone of the great powers, the Russian armies were marginally inferior in both will to aggression and technique. The skills of Russian soldiers won them some very great victories against Austrian armies, but their supply system collapsed, and they broke in exhaustion against German élan. The people and soldiers in 1917 turned on the government that had taken them to this premature, unpopular war. And a new Russian civilization began to be built.

In that same year of Russian collapse, French armies mutinied and soldiers on the march toward the bungled mess that was the attack known as the second battle of the Aisne are said to have made *baa*-ing noises to show they were going to the slaughter. General Pétain made his reputation putting down these mutinies and, by a show of confidence and competence, restored hope to French enlisted men. In the British armies a small portion of officers began to think pacifist thoughts, particularly some who had taken to soldiering with excitement at the start. Siegfried Sassoon, Robert Graves and others were to write down this transformation within themselves with a genius that clutched the thoughts of the whole succeeding generation of the English officer class in books like *Good-bye to All That* and *Memoirs of a Fox-hunting Man*. The shades of these thoughts supported the Chamberlain government at Munich and held back the British will to fight Hitler until it was almost too late. Twenty years after the Somme students at Oxford University voted "that this House would not fight for king and country." Britain had entered the war for her ancient reason of seeking to prevent any continental power from becoming preeminent, the same fear that had sent Marlborough's splendid armies to fight so remotely far from their island. This purpose must have seemed thin to more discerning young men who saw war only as death before machine guns.

German and Austrian soldiers went through this same process of sapping of morale. They were "winning" in that they fought on other countries' lands and one of their enemies, Russia, had vanished. But they had no trick to go on winning. In truth, they were up against that cardinal fact of the European West, that common weapons meant that real aggressive victories were impossible. There could be no military victory; the only hope of prevailing was the collapse of the enemy society, as had happened in Russia. But the British and French were then bolstered by the enormous numbers and productive power of the American nation. And before this reality it was German society that collapsed in revolution.

There was one positive outcome of the First World War. The Germans, at last, found themselves in real need. The Germans had spent all their savings to buy armies. What foreign trade outlets they had owned were taken away from them. Various awkward restrictions were placed on their ability to better themselves by industry, and their military defeat had cowed the vital self-confidence of governors, so that Germany was no longer run by efficient committees. Germans suffered gross inflation, depression, unemployment. Not only were the people given little chance to hope, but many of them were in actual want. All this could, in theory, have been solved by economic and technical ingenuity. The German lands could still provide a good life for all the Germans then living, if properly used. But this argument did not seem very convincing to people living in actual want in that beaten country. Moreover, no one had made false the argument that one day, though it be a long time off, more Germans would need more. All this was seen by a man of ferocious understanding, Adolf Hitler. So vaunting was his confidence in his own view of things, and in his own ability to put things right, that he made himself master of Germany.

In the second paragraph of the first page of *Mein Kampf* appears the following,

When the territory of the Reich embraces all the Germans and finds itself unable to assure them a livelihood, only then can the moral right arise, from the need of the people, to acquire foreign territory. The plough is then the sword; and the tears of war will produce the daily bread for the generations to come.

This passage was written before Hitler came to power and it includes the word, "Only." Only when the people were united and needed more would it be right to take what they wanted by force. But Hitler put the Germans together into one Reich, even the Austrian "Germans" who had never lived under the same government as north Prussians and others since the start of Germanic history. He did this even as he put together the new-style army which he prepared for the next step. The threat of force, the memory of the last war as man against machine gun, and the pacifist feeling that anything was better than war, let him do so.

Soon most of the Germans were in one Reich and one part of Hitler's "only" was behind him, but were the needs of the people so great that the "moral right" arose to acquire foreign territory? We must look at this coldly, without letting Hitler's gross abuse of the words "moral" and "right" force us to deny Hitler's premise out of hand. Hitler's policy was to let Germans live well. By making weapons, paying soldiers, and building roads in the policy of "guns before butter," he dragged Germany out of the Great Depression. These policies did for Germany what Lord Keynes advocated for Britain, and what the Tennessee Valley Authority did for part of the United States. Like them, "guns before butter" was government make-work, which is a shaky framework for lasting prosperity. Even the semblance of prosperity which the Third Reich had reached, with its large Germanic country put together again, seemed to have been possible only through making guns. The next step seemed clear enough, to use the soldiers to take the foreign territory which future expansion would need. Hitler prepared, with a skill which was perhaps the equal of that of Philip of Macedon and Alexander the Great, an aggressive war of conquest; Europe had not seen a real one since Napoleon was beaten, now it would.

Hitler did not aim for the tropical colonies of which nineteeth-century Germans had dreamed, believing that these would be of little use in a crowding world. Germans would not want to settle in the tropics and would be hard to govern if they did. Hitler opted for good continental living space next door. He decided to acquire land in Poland and European Russia, to remove the people he found there and to let the Germans spread all over it. The Greeks and the Romans had done this sort of thing, and modern people thought them rather admirable; now Germany was going to do it.

Such an enterprise required, as it had always done, not so much an enormous army as one which was technically superior to its victim's. The invention of tanks and close-support aircraft let the generals whom Hitler used provide him with the necessary technical superiority. They used tanks and planes in a new array which the British soldier, B. H. Liddell Hart, who had worked out the theory of this technique, called the "expanding torrent." But the German soldiers who made it work found a more expressive slang phrase, "blitzkrieg."

Blitzkrieg was not so much technology as tactics. Decisive though it was against nations without it, the prime ingredients were organization and plan, rather than the tanks and aircraft used to carry out that plan. All the Western powers had tanks and aircraft; it was the way in which they were married to each other and to infantry that mattered. It was this organization which marked the revolution in war.

In the original theory of the "expanding torrent," Liddell Hart had directed his thoughts to the role of infantry as much as to tanks. Modern firearms released the foot soldier from the need to walk with his fellows; he should be an individual, versed in field craft, traveling light; his drill should not be that of the parade-ground ranks but of how to act to orders on his own. Intelligent, self-reliant soldiers would attack by infiltration, perhaps at night, spreading through enemy lines and deep beyond them; advancing where there was little resistance; turning aside where the enemy stood, as the waters of a torrent spread round boulders in their path.

As the torrent of infantry spread on its way there would be confusion in the enemy system of command; communications would be cut; outposts and headquarter units would be surprised; that cloud of worrisome uncertainty known as the "fog of war" would hang deep in the enemy lines. Indeed, his lines would almost cease to exist, as the foot soldiers traveled far and fast into the enemy land. And Liddell Hart thought of some of the men being moved on in lightly armored trucks, leapfrogging their own men who went first to find a way through resistance, rushing the torrent on to flood through the enemy positions.

Adding tanks and close support aircraft to this initial scheme was the obvious next step. The tanks would punch past machine gun nests and be fanning out in the first minutes of an attack. Dive bombers would pick out gun emplacements and roads, pressing the enemy down as the tanks and infantry trucks rolled on. The clogging fog of war would spread an order of magnitude farther and faster than it could be taken by the infantry alone; the torrent would run more swiftly and expand the sooner. Very much of the enemy's force would not even have to be fought at all; it would simply be left behind to be made to surrender later. This final expanding torrent of tanks, trucks and infantry trained to skills in field craft made up the new lightning war that the Germans called "blitzkrieg."

Liddell Hart was a British soldier when he first published these unconventional views. He was retired on half pay at the age of twenty-eight, part of the reason certainly being that he had suffered a damaging wound in France. Another British soldier, J. F. C. Fuller, expanded on the same ideas and got to be major general before he was retired. A few others of the younger generation in the British army read and listened, but the high-ranking enthusiasts for the new idea were German. A Colonel Heinz Guderian read one of Liddell Hart's essays, saw an experimental tank brigade of Fuller's, went home, got the highest possible encouragement, and founded the German armored divisions on what he had seen. One of Liddell Hart's books was translated into German and became a textbook for the Waffen SS. But young German soldiers were thinking on the same lines, anyway. A junior officer in the first war, a certain Erwin Rommel, wrote a book called *The Infantry in Attack,* expressing many of the same ideas; he was given one of Guderian's new Panzer divisions to command. And similar thoughts were coming from the younger generation elsewhere, for instance, in a little book called *The Army of the Future,* published in 1934 by one Charles de Gaulle. Similar thoughts ran through the mind of a young American called Patton, and some of his countrymen, though these soldiers were not in the European calculations. And many Russians too, as the world found out in time. But it was in Germany that a high command believed the new thing and acted accordingly. When Fuller was invited to Berlin after his retirement to see a military tank parade, Hitler came over to him and said, "Well, what do you think of your children?"

It was on this technology of blitzkrieg tactics that Hitler's Germans had to rely for success in aggressive war. The armored divisions and tanks were symbols of the new array, but the real trick was the coordination of tanks, infantry, aircraft and supply. Success would come from training and arrangement, not from hardware alone. At the start of the war only Germany would have a blitzkrieg army and would gain rapid victories. But their victims, and their victims' friends, would learn the new thing very quickly. If the war were to last much more than a year after the first showing of an expanding torrent, German soldiers could expect that their future enemies would be organized in the same blitzkrieg array.

To understand the course of Hitler's war, therefore, it is necessary to know three things. Firstly, the war was a deliberate aggression aimed at securing permanent niche-space for rising numbers of a comparatively affluent people; it was not, therefore, an attempt to gain dominion over Europe, though that might be a byproduct. Secondly, the target was Russia, with the object of expelling Slavic peoples from the East European plain and taking the land for Germans. Thirdly, the military technique prepared for the conquest was vulnerable to imitation, meaning that the conquest of Russian lands must be achieved very shortly after the start of the war.

Hitler attacked Russia more or less on schedule in 1941. But first he got himself involved in war with France and Britain in 1939, an annoying accident from Hitler's point of view because it made him show the superiority of the new tactics before he was ready to use them against Russia. They turned out to give results as spectacular as any of their proponents had imagined, and the expanding torrent of German armies rushed through France in just six weeks, leaving scattered French soldiers no option but to surrender. The French had actually had more tanks and other modern hardware than the Germans, but it did them no good against the new array.

Then there came an incident that puzzled many. The small British army was cornered against the sea at Dunkirk and commanders of German armored divisions were preparing what they intended to be the massive assault which would compel the British to surrender. But Hitler called them off, forbidding the panzer divisions to attack. And the British soldiers crossed the Channel in a fleet of small boats. The German commanders were annoyed, yet Hitler's purpose is clear enough and his decision typical of his wisdom. There could be no expanding torrent against the British army turned at bay; it would be a little war of attrition. True, the British would certainly be destroyed, and quickly, their tanks going down before superior numbers, their field guns rushed by armor as they fought it out over open sights, their infantry rounded up and forced to surrender, but the Germans would pay a definite price. Hitler had seen what attrition was like himself, twenty years before. He would lose tanks and men he wanted for the attack on Russia: "Let the British swim away, they cannot do us more harm." It was a reasonable attitude.

Then Hitler was in a predicament. He found himself with near dominion over Europe, but he had not yet struck at Russia. He had the biproduct of the planned aggression without the product itself, and yet he had revealed what the German armies could do, and how they intended to do it. Russia must be conquered before Germany was faced with powerful armies wielding the same technique. The need for a quick strike was really pressing and so Hitler abandoned an obviously futile plan to float his blitzkrieg on flat-bottomed boats to England, sent his soldiers to the east instead, and attacked Russia. People often wonder at the apparent foolishness of adding Russia to his enemies. A miscalculation it certainly was, but it wasn't foolish. The main purpose of Hitler's armament had all along been an aggression against Russia to appropriate some of her territory. To give this endeavor a reasonable chance of success it was necessary to strike while the German armies still had a technical edge.

Russia was saved because her leaders had been reading the Western military textbooks in good time. The young German officers of the thirties were but half of Liddell Hart's readership; the other half, as his own memoirs clearly show, were the revolutionary soldiers of the Red Army. In 1941 the Russians could field twenty thousand tanks together with close support aircraft of their own, and they even had in their multiple rocket launcher (the so-called "Stalin Organ") a science-fiction blitzkrieg weapon superior even to those of the Germans.

It took the Russians time to learn to use their new weapons (partly because Stalin had just shot all the senior officers), and they had trouble matching the literate initiative of the German command system. But they had the best of teachers in their first desperate months of defense, and then their old ally, General Winter, gave them time to profit by what they had learned. From then on neither army had a significant advantage in technique or skill, and the issue was decided by the simple fact that there were many more Russians.

In the end, of course, the Germans also had the "second front" armies of the Americans and British to contend with too. These second-front campaigns give us instructive lessons on what happens when two armies in blitzkrieg array meet head on. Something like the old days of Marlborough's and Napoleon's wars returned, where the issue of battle turned on managing to hold down as large a part of your enemy's army as practicable before launching your rushing torrent at a weakened front. The Normandy landings were like this, as a long grinding fight by Montgomery's Canadians and British over Caen dragged German reserves into one flank, just as Marlborough's assault had dragged the French reserves into Blenheim village. Then Patton unleashed his torrent of tanks on the other flank, swept through and around the whole German army and compelled it to surrender in the pocket of Fallaise. But all this is interesting only to the student of battles. The actual issue of war was decided when it was found that Hitler had not the technical means to beat the Russian army.

Hitler's attack on Russia was the most thoroughgoing attempt at real aggression which Europe had seen since the Gothic barbarians removed the Romans and set up their own kingdoms. Even the Moslem and nomad invaders of distant centuries were not quite so pure in their aggressive intent, for they always had half a mind to go home with plunder rather than to settle. But Hitler's Germans meant to take the land as a Roman legion had meant to take it. The German failure confirmed, however, that in a literate community the technical superiority necessary for such an endeavor could not be sustained.

Japan

T HE JAPANESE learned of the ways of the European West when their island was feudal, agrarian and crowded. So crowded were they that, although a country of rural folk and artisans, Japan actually imported its staple food, rice. Her feudal masters kept their country and their own power isolated from the expanding West as long as possible, but the traders came at last. The Americans were first, extracting a trade treaty with a few thoughtful discharges of their cannon, and the Europeans were glad enough to follow where the American cannon went.

It took the Japanese less than ten years to act on the politic maxim, "If you can't beat them join them." In a mild revolution the feudal chieftains returned their fiefs; their quaint old retainers, who served as soldiers, were pensioned off, and the people turned their formidable brains to mastering the techniques of Western industry. They also bought or manufactured some cannon as they went along, because they were a trading nation and they had learned from the civilized West that cannon were a very useful aid to trade.

The people learned the new ways in a single generation. Their aspirations bounded high, and so did the country's birth rate. Between the revolution of 1868 and the turn of the century the Japanese bred an extra thirteen million people. They were so crowded to start with that they had to import rice. Now, as an industrial state, they had to import raw materials too. The people felt that they needed to be sure of their supply of resources; they needed to own land with raw materials in it, to control outlets for some of their trade and, perhaps, to have some land where they could deposit surplus people too. They prepared to take what they wanted by force. They studied Western war, made what Western weapons they could, and bought the rest. Britain had the best navy, so the Japanese bought British warships. Germany had Krupp, so they bought German artillery, and so on. Then they beat up some primitive Chinese forces to let them have their way over some minor matters on the mainland, which gave them a chance to try their hands at the new form of soldiering. Practice was necessary because they knew that they must tackle Western power eventually.

What the Japanese wanted first was Korea, a long-coveted piece of land which could supply many of their immediate wants. Curious as it may sound, they had to fight Russia for the privilege of conquering Korea, because the Russians felt their vital interest was involved. So they fought the Russians and beat them handsomely in a campaign of 1904, known as the Siege of Port Arthur. They had read the very latest military writers more carefully than had the Russians, and were that much more modern. The Russians let them have Korea. It was a very successful aggression.

But, as the population went on growing so did the needs of the people. The Japanese got some more land out of the First World War when, by a very sensible arrangement for mutual benefit, they were allied to the British. Their purchase of cannon and warships was paying off very nicely. No other Asiatic power stood a chance against Western methods of war, whether these were used by Japanese or by the Europeans who had invented them; and the Japanese had carefully avoided a clash with a major Western power. Even in their battle with the Russians they had prudently kept themselves to very limited objectives, avoiding such provocation that the Russians might think a major war effort was necessary.

But aggressive war was now a Japanese habit, as it is to all nations who pursue it successfully. And the Japanese numbers and aspirations still continued to grow.

Some of the thinking then common in Japan was written down for us by a young Japanese naval officer in a book published in 1935 and called *Japan Must Fight Britain*. The theme is that Japan must expand; therefore, someone must move over. If not, "Japan must fight Britain," and if Japan loses that first fight "It is as clear as day that, with her population and her insufficient resources, it would not be long before she had to draw her sword and stand up to fight for her life."

The Japanese were convinced that they must take by force what was needed to give an ever-growing population the standard of life it thought it deserved. And they did fight. In choosing Britain for an enemy they were, of course, tackling a weakened power, one with incompetent generals who had lately preferred horses to tanks and who had not thought it necessary for Singapore to have guns on the land side as well as on the sea side. Even so, it was a dangerous thing to do. But tackling the United States of America at the same time was so hazardous an undertaking as to be scarcely believable. The reasons for this desperate venture must have been very strong indeed.

A wealthy island state, as Japan then was, is predicted by the ecological hypothesis to be extremely prone to start an aggressive war. The opportunities for a broad niche on the island base must be limited, putting a strong pressure on those desiring wealth to find opportunity elsewhere. Trade must, and always does, become part of the niche-space of the affluent on all islands that achieve moderate wealth. But part of the success in winning niche-space through trade must be spent, through the breeding strategy, in the production of more potential traders. The numbers of the affluent grow, but there is also a worse population consequence as well because imports always include cheap food, which allows the poorer classes to maintain their own, larger breeding effort. The country then becomes dependent on free access to markets to maintain its traditional affluence and must be ready to fight for those markets. A few generations later rising numbers of the wealthier classes find it difficult to provide affluence for their own descendants, which is the essential condition for an aggressive war. Island folk can educate, trade, and breed their way into this condition very easily.

It was clear to the Japanese of 1941 that they had real need of access to other people's land if their new and better way of life was to be maintained. They were quite right. They still need that access forty years later, and are getting it. The large continents of Europe and America are yielding to the Japanese much of their continental living space, welcoming Japanese traders, keeping the sea lanes safe for Japanese commerce with their own navies, accepting Japanese manufactures even at the cost of destroying their own currencies and industries. It is, perhaps, a moot point how long they will feel content to go on doing so. But in 1941 the Japanese were not being so warmly welcomed in the world.

The Japanese had long been fighting to take provinces from China, expanding the continental land they already held in Korea. It was, of course, blatant aggression, sired of ecological necessity and mothered in imitation of the Anglo-Saxon example. Success in this war, long and drawn-out as it came to be, was vital to those who controlled the Japanese government. The people too were deeply conscious of their need, schooled to accept the new Western ways, building a high standard of living by fighting for it. And they were proud.

Then the United States did two things for Japan.

They placed an embargo on the sale of the oil needed to sustain Japanese armies, and they put all American warships into one handy disposable package in a harbor in Hawaii. A better combination of stick and carrot to drive the Japanese into war would be hard to imagine. The war chiefs saw their opportunity and struck, and Japan was committed to the forlorn, hopeless adventure. They would take their oil from the British in Burma, by force. They did that. Then they would use valor and skill to make the United States agree to their keeping the loot. But this could not be.

England

AN ISLAND with a longer and bloodier history of aggression than Japan is Britain as controlled by the English. The English have taken some portion of their daily living through the trade of aggressive war ever since Norman Duke William consolidated the kingdom nine hundred years ago. They conquered Wales and made it a principality of the English king. They fought the Scots for centuries, though their armies were always held at the edge of the Highlands. Finally Scots and English compromised by sharing a king, thereafter to go aplundering together. And they fought to expand into France, long and desperately in those fights called the Hundred Years' War. They fought as archers

300

in free companies, mercenary and plundering men trained to use the longbow since childhood, brought up with the idea that they would get their living with weapons. Or they fought as gentry in armor, living in the niche-space of military command in France, seeking ransom as a prime goal of existence or, for the most powerful and lucky, a French estate over which to be made lord. Soldiering gave hope for betterment as no other life in England did, and fighting in France was to the English of those days what going west was to young Americans of the nineteenth century.

Like all aggressors, the English had a technical advantage in war which made fighting an attractive source of livelihood. They used longbows to shoot down armored horsemen, massing their bowmen in dense companies amid thickets of sharpened stakes where horsemen could not follow. Then they mixed these human fortresses of longbowmen with squadrons of armored horsemen to prevent enemy infantry from closing with the bowmen. This was vital because an archer had neither armor nor heavy personal weapons, needing to be able to move freely in order to use his bow. English longbowmen could dominate a battlefield only if they were protected, rather as field artillery have had to be protected in later wars. But when the mix of bowmen and knights was skillfully done the effect on an attacking force in a pitched battle was dreadful. At Crécy, Poictiers and Agincourt, well-armed Frenchmen, fighting with hard determination and high courage in defense of their own land, went down before the storms of arrows in such numbers that it seemed beyond all reason. Courage is never any use against superior technical means and it is not bravery that gives the victory in war.

Du Guesclin and Joan of Arc found the technical answer to companies of longbowmen. You did not hurl your knights against thickets of stakes hiding archers, you held off, waited for them to get hungry, and harried them on the march. You built fortresses, held together as strong points in a unified system of command. You denied battle except on grounds of your own choosing. The longbowmen were now useless and the English aggression collapsed, swept away in a few years so that Frenchmen ruled all of France again and England's frustrated soldiers were tumbled back into England to try to live without hopes of loot.

They took to fighting among themselves in a long civil war called Wars of the Roses, until a captain strong and ruthless enough to impose his will emerged, the familiar and inevitable pattern when countries full of ambitious soldiers engage in civil strife: Philip to be king of Greece, Caesar to be dictator of Rome, Genghis to be Khan on the steppes. It was Henry Tudor to be king of England, to control the country tightly by force, to rule through informers and secret police, and to have special courts for trials of political prisoners, the Courts of Star Chamber.

Henry VII, Henry VIII and Elizabeth, the powerful Tudor monarchs, provided law and order for the English, but also the hope that comes from growing resources. They found, in the encouragement of industry, scientific agriculture and trade, a policy so agreeable to the English that it consolidated their personal power. Subjects who wanted to fight their way to fortune could do it by fitting out ships to loot Spanish galleons coming home with treasure stolen from Aztec and Inca civilizations in Central and South America, or by attacking Spanish settlements in America directly. Subjects who wanted money from their land could win acts of Parliament that "enclosed it," which meant that they threw peasants off land they had cultivated for centuries in order to raise sheep for wool, or grain for the growing cities. The English countryside took on in Tudor times that look of small fields with trees and hedgerows that it still has, and many of the handsome Tudor farmhouses, built with the new wealth from agriculture, still stand. The cities, particularly London, grew rapidly with the influx from the countryside, the new arrivals finding work there as artisans or laborers in the new industries.

As in faraway Sweden, the Christian Reformation was suited to a society bent on changing its economic ways, but wise and heartless Henry VIII found a very special use for it—quite apart from his manipulations of the Church to give him a succession of wives. He used the Reformation as an excuse to loot the monasteries for capital with which to run his exuberant police state. The Church had grown wealthy in capital and rents, owning gold and silver in coin or ornaments, lead on its roofs, and prize farm land. Henry "dissolved" the monasteries. He sent a small party of armed men to each, requiring that the prelate in charge hand over his gold, silver, lead roof (which regrettably meant destroying his buildings) and the title to his lands. If the abbot yielded gracefully he got an adequate pension to keep him in comfort for the rest of his life, if he refused he might be hanged in his own courtyard or otherwise subjected to unpleasantness. Most yielded readily enough because the cause of a wealthy and opulent church as lord of all manors was not popular in this ambitious England. Henry paid for his government with this confiscation of church property, finding it a more popular way of raising money than through taxes. He left entrepreneurs alone with their profits to reinvest, and his church loot provided a sudden influx of new capital into the whole state. At the end, though, the church money gave out and Henry was reduced to debasing the coinage to pay government expenses, like other despots before him.

This plunder of the Church for industrial capital and tax money also excused a rapid dismantling of the ancient social order very similar to that achieved in nineteenth-century Japan. A squirearchy replaced the Church in rural government, much of it absentee and controlled from the one center of power, London. Affluent people in Tudor society were fond of drawing attention to their success with references to "my manor at ———" These manors used to belong to the church.

The English population had been growing steadily since the great cull of the Black Death in 1348–49, but it probably grew particularly rapidly in the century of Tudor rule until it reached the apparent density of which Richard Eburne complained in his *Plain Pathway to Plantations.** As Elizabeth Tudor died in 1603 they were beginning that emigration and expansion into the stone-age parts of the world, and to obsolete Asian empires, that would give them the largest conquered empire in history. Their numbers, their ambitions, and their industrial success had grown together. At such times the ecological hypothesis states that aggressive war against poor neighbors, or social oppression, or both, will happen. In the Tudor police state a new, controlled, and tightly organized society had evolved, giving that new social contract of master and man which was the first niche carried to New England in America and which the settlers were to fight over two hundred years later. The land of England was made into a productive

* See page 220.

garden, owned by landlords, farmed by tenants, producing in the service of city government. The Tudors had personal objections to large armies and martial affairs, because their own power was secure only after Henry VII had disbanded the rival forces left over from the long civil wars, but they engaged in aggressive war against their poorer neighbors all the same. They conquered Ireland, which was almost defenseless against contemporary armaments, but to such poor purpose that the fight still lingers on four hundred years later. And they attacked Scotland, though once again to see English armies beaten off by the dour defenders of the mountains.

Yet the grip of despotism relaxed slowly through the Tudor century, which was a measure of success with the other responses to rising numbers and ambition. Capital stolen from the monasteries had fueled economic expansion, and grand agricultural and engineering projects had been undertaken by the state and through deliberate acts of Parliament. The king, and the new wealthy around him, hired engineers to drain swamps and build businesses, being quite ready to purchase these skills abroad, often from Holland, which sent drainage experts and know-how for the new crop rotations. They were acting as Japan would act three hundred years later, the people of an island needing to catch up in technology with larger continental powers, and not being too proud to seek advice in return for hard cash. These things, and the new order in society, meant that wealth managed to keep pace with ambition and numbers for a while so that the police grip of the Tudors could slacken. By Elizabeth's time the English could talk much about English liberty as they directed their ideas of free enterprise toward the New World.

The new economy and social order would go on providing more and more resources for the English for a long time, letting many more people be supported in the island even in a happy state, let alone subsistence. Eburne overlooked this when he said the population must be reduced by emigration and at once. But his insight was penetrating all the same and his advice some of the soundest on which the policy of a state has ever been based.

Grandeur in a people's individual hopes depends on a large surplus of resources so that there need be no strictness in the rationing of them. It would have been possible for Eburne's England to have forgone emigration and foreign conquest, and to have let her population go on growing for a while, but soon her people would have found those individual happy dreams obstructed by bureaucratic law and bureaucratic order, both necessary to a strict sharing process. A people so lately helped to ambition would have resisted this, some of them defending their own expanded way of life by seeking the privilege of caste or rank, and others by rebelling against the laws of privilege in the name of their recent liberties. The people would have come, at last, to some compromise with life, but the story of England would no longer have been interesting. It would have been like the story of Sweden after the failure of her aggressions in the early seventeen hundreds.

England chose emigration. To win land for this emigration she undertook the most successful aggression in the history of the world, using the weapons of the European West to take the North American continent from the Amerinds, and then Australia, New Zealand and smaller countries from their original inhabitants as well. The North American conquest was completed by her successor, the United States; the rest she did all on her own. Her aggressive conquests gave her not only a place to send surplus people, but also supplies of the raw materials of manufacture and outlets for trade, which should maintain the impetus of the economic expansion at home. She then multiplied these trade outlets by conquering populous Oriental countries not provided with the war techniques of the European West, and using both their raw materials and their peoples as part of the niche-space needed by the growing British aspirations.

The only technical difficulty in these aggressive conquests was offered by rival European powers. Britain had to fight over the spoils. Geography gave her a great advantage for this, and she made the most of it. By her wise policy of backing the weaker side in all European disputes she made sure that no European power should so dominate the continent that European energies could be concentrated on disputing her rights to colonial loot. She thus never let herself play Carthage to a European Rome. And, as an island nation, she was able to make for herself the most efficient fighting fleet. She was never able to obtain a technical advantage in the design of ships or their weapons, but through much practice in ships always at sea she was able to get more out of contemporary technology than could her rivals. In 1805 a British warship fired three broadsides to a Frenchman's two. This was enough. It led, through many petty battles and such set pieces of generalship as Trafalgar, to the largest wholly owned empire the world is ever likely to see.

To this glut of imperial land was added the greater glut of industrial innovation. The British became the first to use thermal energy to run factories. Even as the French Napoleonic empire rose and fell in Europe, the British made steam engines, power looms, and even the first steam carriages. These provided her with enough income to fight Napoleon without hurting her capital. In 1818 a British admiral said in the House of Commons that the war with France would have brought England to "total ruin" but "for the timely intervention of machinery." This extraordinary happening of the industrial revolution let the island trading state not only survive a long war of attrition against a continental power but also to emerge from the war wealthier than when it had started. This seems to be without parallel in history.

The British responded to their good fortune in the expected ways, with large families and cries for "liberty." They tried to get rid of the results of the large families by shipping the products to the new lands, but still the population grew. Soon their island was so crowded that contemporary agriculture was hard put to feed the people. Rationing by price brought notice of this stern reality to the huddled masses in the towns. But the sense of plenty which the new ways ought to bring had already brought its quota of pressures for "liberty," which the British governors had answered with an ever-growing electoral franchise. The new electorate voted for what seemed the obvious way out of their population dilemma—abandoning the attempt to grow their own food and fetching food from the imperial lands they owned or from the United States of America. The "corn laws" (tariffs on imported food) were repealed. From henceforth the ever-growing population of the island of Britain would be living, not just on the island rock on which they had their houses, but on great continental spaces beyond the seas where lay most of the other parameters of their niches.

Before the corn laws were repealed, English agriculture was in an extremely healthy state. One can well imagine that it would be, for large and bounding populations were fed from Britain's very fertile island. The English half of Britain has a climate that is close to ideal for many crops and pasture, with weather that is never very cold, never very hot, and where rain usually falls every month of the year. Tourists may not like this, but grass, cattle, wheat, barley and potatoes like it very well. Farmers grew rich, and made a fetish of the excellent state of their fields. From this period comes the image of ruddy-cheeked, plump John Bull, the English countryman incarnate. Elaborate crop rotations were practiced and there was a heavy use of fertilizer; not the chemical kind, which was not around then, but horse manure hauled out of the stables in the big cities by train. British animal breeders of those days gave us the Dairy Shorthorn, Hereford beef-cattle, and many of the sheep and pig breeds we still use. It was farming to gladden the heart of a modern advocate of organic foods, and it paid very well.

But grain taken from virgin acres in America could be sold in London for far less than it cost to grow grain on an English farm, where the land rent was high and yields could be kept up only by paying for all that horse manure. American farmers stole the land from the Indians, put back no fertilizer, and accepted low yields because the acres to be used were almost unlimited. Their products were cheap. Later, when refrigerated ships were invented, the same cold logic meant that meat raised in New Zealand, Argentina and Australia could be sold in London for less than it cost an English farmer to raise it; and not all his skill, even with the best sheep pastures in the world, could do anything to equal the price of meat raised on distant looted lands.

The high quality British farming was kept going only

by the tariffs of those hated corn laws. But more and more people wanted cheap food, they had the vote, and the corn laws were repealed. It meant, of course, the complete ruin of British agriculture. Much prime farm land was simply abandoned so that a large part of the country districts of England became left to the secondary plant successions, to brambles, thickets, and the roughest sort of pasture. This gave entertainment to townsfolk, for places like the South Downs were full of wild flowers and lovely to roam in. There was good fox hunting to be had, and rough shooting. But there was a low and poverty-stricken rural population left behind. Like the Romans of the old empire, the English had decided to feed their cities on distant real estate won for their use by superior weapons, and to let their own land go empty.

It was a perilous decision which the British had made. For as far into the future as they could see their welfare was to depend on their free access to the fruits of distant real estate. Suppose other peoples laid successful claim against that distant real estate? Suppose the indigenous peoples so multiplied that they needed all the fruits of their land for themselves? Suppose the martial British grip relaxed so that her people were no longer willing to hold down distant rebellious subjects by force? All these were possible; indeed, they were all to come.

But the more immediate British peril was that some rival European power would interfere with her trade routes. So the British set about building and maintaining a navy so large that it could overwhelm the combined navies of any conceivable alliance made against her. For a century this navy patrolled the sea lanes of the empire. So visible were the British fighting ships that they were given credit for the general peace which prevailed among the greater powers, and people talked of the "pax Britannica." In truth it was not the British navy that gave the world peace but the surplus niche-space available to the European West. The industrial revolution and the empty lands of the Americas acted as siphons for the surplus people and the surplus hopes. Until these niche-spaces should be filled there would be little impetus to aggressive war. It took only a century for the breeding efforts of Europeans to fill these niche-spaces with people, after which the falsely named "pax Britannica" ended and the great wars came back.

The sense of omnipotence at sea which the "fleet in being" gave the British was always illusory. They were left to parade their warships only as long as others did not really need what they had. Once this happened, others would build fleets too. Anyone can make a war-ship, just as anyone can make a musket, a cannon or a tank. By the end of the nineteenth century, when both Europe and America were showing signs of filling up, other countries did start making warships. The British omnipotence at sea was at once over and she was soon to use her fleets, not to keep the peace but to fight for her life. The island home survived those fights only because Britain had allies with more capital, more soldiers, more physical resources and, in the end, even more ships than herself.

But the British Empire was lost in all but name even before the mighty fights with Germany and Japan stripped the British of their capital. All the imperial spaces began to fill with people, and this was true whether their populations were formed from British immigrants or from their own indigenous peoples. Both sorts of population were fast acquiring the techniques and learning of the European West; indeed, it was the humane British policy that they should do so. Soon their rising numbers and rising aspirations would make them all divert the bulk of their resources to their own use. With aspirations as high as those of the British themselves, they would soon want to run their own affairs. And once they had access to the weapons of the European West there would be no way of denying them this wish.

In the event, most of the imperial peoples did not have to fight to be free of Britain. Those of British stock were let go early with a parent's blessing, in the fond hope that they would look after the parent's needs thereafter. The turn of the remaining countries had not yet come when the wars with Germany and Japan so depleted the British treasure that fighting to retain an empire did not seem to the British a proper thing to do. But it is quite certain that, even had Britain still the appetite and the capital to attempt to hold on to that largest empire in the history of the world, by force, she could not long have done so. She would have been opposed by multitudes of people equipped with modern firearms. A machine pistol makes every man and woman a useful soldier. There would have been just too many people carrying machine pistols for the soldiers of a small island power to suppress.

Then the British found themselves a crowded people, on a small island, with diminished resources. Their industrial supremacy had gone, for the world had copied them or even passed them by. They had no empire to use as niche-space, either carried to the island in British bottoms for trade or used *in situ* as adventurous youth and upper classes governed it. They had to buy their daily bread from the proceeds of trade carried out along sea lanes kept peaceful by the navies of other powers. This was the future promised only a hundred years before when they repealed the corn laws and decided to multiply and live by other peoples' land. It had come very quickly.

Europe never fell before military conquest; its weapons, its learning, and its neighbors saw to that. But Europe is coming very close to being a single political entity all the same. The superstate, which is emerging after all the aggressions failed, is called not an "empire" but a "common market."

Up to this final act of union the pattern of change and war in Europe has, in fact, followed very closely the predictions of the ecological hypothesis, though on a grander scale than in any previous civilization. But the scale has been a product of European technique and learning, just like the wars. The historic pattern itself has been familiar enough.

In each part of Europe niches have changed, as many people learned new techniques and began to live better. Always the breeding strategy responded to produce more people. Except in the last century, when fossil-fuel technology so increased total niche-space that broad niches have spread down the social pyramid, the absolute numbers of poor people increased. Growing aspirations and numbers of the better-off in different European nations have always been accompanied by trade and colonial expansion, as is to be expected if the one causes the other.

All the European states have undertaken aggressive wars against distant peoples not equipped with weapons of the European West, and have taken, settled and used the lands they took by force. All the European states have attempted aggressions against the members of their own polity, and always at the times when the aggressors were comparatively wealthy and had a high standard of living. These wars were most persistent and vigorous in the last and wealthiest century, when absolute poverty was in decline. And now, at last, the European nations are moving toward unity with a central government, quite as predicted by the ecological hypothesis, but without the complete military conquest that usually imposes the unity. But the way to unity was probably prepared by the tries for military dominance all the same.

Central government, bigness of scale, and fusion of national identities into a large communal state are themselves ways of offering fresh opportunities to traders, governors and the more aspiring members of society generally. As long as the new supergovernment is not hostile to your own personal interest it is almost certain to bring more opportunity to you and yours. There will be more opportunities for trade, more niche dimensions in travel, a greater variety of resources on which to draw, and more military security too, within an amalgam of states. This is why an imperial conqueror, if not an expropriating alien, can be accepted. Europe never had a conqueror, but ancient Greece did, and the Greek example is instructive.

When Philip of Macedon ended the centuries of Greek civil war with his superphalanx, the wars really did end. The people of all the Greek city states accepted the new system of things, and they were soon able to rejoice in Alexander's conquests in their name, and to take grateful advantage of the opportunities his conquests gave. It was Philip rather than any other soldier, who imposed unity, because he it was who invented the winning technique. But it seems likely that any of the other would-be conquerors, from Athens, Sparta or

Thebes, would have done as well had they managed the necessary victories. It was, in fact, possible for all the Greek city states to share in the advantages of Greek union, whatever the name of the soldier who imposed it.

In Europe no soldier managed to impose union. Furthermore, the more advanced technology of Europe had let a society develop with much wider individual liberties than most of the Greeks could ever know, and the idea of soldierly government has been resisted strenuously by Europeans. But this has not obscured the advantages that could come to the better-living individuals in the merging of the nation-states. Both numbers and aspirations have grown rapidly, meaning that more carrying capacity must be found. There are more broad niches available in the superstate, and so people have found a way to build that state without the necessity for a successful conquest. The nations joined by free votes of their peoples, each vote being given not on the issue of defense or grandeur, but on economic logic: "the market will improve my standard of living." The Common Market is a consummation of the expansive phase of the cycle of European history, the equivalent of the old land empires of continental scale and, like them, built after much fighting.

The Common Market has even let the British in. For seven or eight hundred years British armies have fought their way into Europe with never a continental army getting back to Britain in return. In the early days the English soldiers were blatantly after the loot of European lands, as their appetites and numbers began to stretch beyond the satisfactions of their island. In later centuries it was soldiers of a united Britain fighting to prevent any European army from getting so powerful that they could interfere with Britain's martial growth through colonial expansion outside Europe. But the soldiers out of Britain always took war to some European countryside in the service of their own relentless expansion and demand. They fought down European peoples so that the British share of the world's niche-space might grow. They made a great success of it. And now the Europeans have invited the British in to share their continental land after all. This invitation came just after the British had lost the loot of their world adventures and found themselves on a small island which could not support such numbers of them living in the expansive style of the European West. The decision may even have prevented their having to think of going to war again for a living in the coming generations. Most timely.

The need for very great variety in the resources for the human niche of European peoples is now most pressing. Both the numbers and the densities of the people are very large, and the habit of living in broad niches has spread to a large part of them. All must have opportunity, variety, and choice of resources if they are to believe the rhetoric about Western liberties for long. This new European union of the Common Market probably offers as good a chance of meeting so many needs as any earthly polity. It has been built, not out of conformity and by force, but out of variety desperately defended. Different languages and different cultures are all intact. De Gaulle said it was, and was to be, "*Europe des Patries.*" This vision speaks to keeping the precious diversity that all may share it. In variety there are compounded possibilities for free living and broad niches. *Europe des Patries* gives a little extra niche-space to support European hopes and breeding strategies for another short step into the future.

CHAPTER NINE

THE SHAPE OF THINGS TO COME

THE CAUSES of history have not changed. Some parts of the future are, therefore, predictable.

The human niche remains a niche that is learned. As individuals grow up, they learn how to live from parents, from those with whom they mix socially, from the culture of their times. Each generation can acquire some of what is new, so that some of the dimensions of the prevailing niche can change with the tempo of generations. But the ability to learn the new is balanced by a tendency to move slowly in abandoning the old. This conservatism is a very necessary human quality. Like the fixed habits of an animal species, it serves to keep the behavior of individuals within reasonable distance of the tried and tested; it gives them a good chance of living to raise children because it directs them to do what has already been found to be successful; it gives *fitness*.

The human breeding strategy remains what it has always been. Each breeding pair acts to maximize *fitness,* which we define as the number of offspring who survive them to breed in the next generation. Fitness in human breeding is largest when the chosen family is at an optimum, not too large and not too small. But this optimum number is very sensitive to the broadness of the niche to which the children are to be raised. The relatively poor will always have larger families than the relatively rich. The experience of history is that the average family that results is more than is needed to replace the parents, even among the affluent. The only circumstance in which families fall below replacement is in the more extreme forms of poverty, where resources are so constrained that the optimum number falls below two.

In very dense communities, where there is much extreme poverty, the population may be stabilized by a low breeding effort. All other human populations always tend to rise, unless kept stationary by cultural devices to prevent adolescents from being recruited into the adult breeding pool. Taboo and tribal war served this purpose for our remote ancestors but there are few modern equivalents. We can expect, therefore, populations to continue to grow in all countries.

Populations tend to rise most quickly following a large increase in resources or standards of living brought on by a major technical advance or a successful aggression. This is because the optimum family then can be seen to be large by people of most standards of affluence, but particularly by those being recruited from the poor to the middle classes. The spurt in numbers always ends when the new resources, won by technique or conquest, are used up; after which the population continues to increase, but more slowly. Many modern nations have just passed through, or are still in, one of these periods of rapidly increasing numbers.

There is an important variant on the effect of fresh resources on the optimum family. It is that hope, alone and by itself, will raise the number of children chosen. Any reason for rising hope in a population always leads, therefore, to rising numbers. The new hope may reflect some simple influx of new resources, such as the fruits of conquest or green-revolution food, or it may rest on no more than learning that improvements are possible in theory. But hope itself will lead to larger families. This is inherent in our breeding strategy and the prediction is confirmed by the historical record. A feeling of well-being makes the numbers rise.

Rising numbers must always soak up spare resources by sharing them out among the extra people. One consequence of this is that poverty always persists. A second consequence is that good times for the not-so-poor must always end in some successor generation producing a predictable series of events which include trade, colonialism, class repression and aggressive war. Since our own numbers will continue to grow, it is inevitable that our own future holds variants on these themes.

Three kinds of assertion have been made in recent years suggesting that our population growth will stop by itself from some inner dynamic of its own. These assertions have had very wide publicity, for they seem to say that the "population problem" will go away. History of the uncomfortable kind that I have been tracing would then end shortly and my predictions that we are in for more of the same would be wrong. The thinking behind all three assertions, however, is in grave error.

The three erroneous assertions are that populations stop growing as people become wealthy; that the recent "explosive" growth in the world population has been due to medical advances and will go away as people adapt; and that there are signs that the human population is now at the "inflection point," at which numbers will level off as in other kinds of animal, remaining comfortably stationary thereafter. All three of these assertions violate scientific principles and assume that magic is at work in the control of numbers of all living things.

The assertion that spreading wealth will halt the growing populations is a statement of what is called in the textbooks the "theory of the demographic transition." I have examined this fully in an earlier chapter. The idea does not have the status of a formal theory, in spite of the name given to it. It is merely the observation, now commonly made and well established, that more-affluent people have smaller families than poorer people. This is explained by niche theory, which truly is a theory and which explains the observation. There is *no* evidence that making people wealthy will halt population growth, merely that growth will be somewhat slower when we are all wealthy.

The way in which the demographic transition argument is often offered makes it particularly dangerous to the human well-being. In its most glib form it slides out as a sentence something like this, "We now know that poverty is a *cause* of population growth and not a consequence." The implication is that, if we will only get down to producing wealth and sharing it with the poor, history will go away. But that glib sentence is utterly false. It is based on nothing other than the belief that there is some magic in being wealthy that sets the family size at replacement. A rising population *is the cause* of increased poverty; niche theory predicts that it will be so; the historical record shows that it always has been so.

The medical-advances argument I have also discussed earlier, when introducing the European West. This argument asserts that the very rapid population growths of recent centuries are due, not to biological properties of people, but to advances in sanitation and medicine that kept alive children who, in previous times, would have died. Some of the more striking bits of modern history, therefore, would appear to be caused by the medical profession.

The argument has a plausible ring to it. "There are all these extra people because we stopped the children dying, don't you see." But human populations, of both city and countryside, have had many periods of rapid growth before the advent of medicine and sanitation. The peopling of the Americas by Europeans was far advanced before medicine and sanitation arrived, and so was Asia crowded with peasants. The only actual data we have in support of the assertion is that the arrival of modern medicine was synchronous with part of one episode of rapid population growth—essentially the last three decades in the tropics. Strange to say, modern medicine and sanitation are symptoms of that wealth which, on the argument of the demographic transition, is supposed to halt population growth.

For a few years when first introduced, medical improvements probably do cause a few more children to be raised in a single generation, because, as I have said earlier, the families of that generation will have been conceived in ignorance of the effects of the new medicine. The effect has no long-term significance, except to let people plan their optimum family with greater precision. But to assume that the recent invention of mass medicine has made any fundamental difference to the number of children raised in any contemporary society is to assign to people the small-egg gambit of a mosquito; it is to assume that women are mere baby factories and their output is a function of what the doctors can keep alive. It is unscientific as well as literally inhuman.

This leaves the third assertion which can best be described as "the doctrine of inflection in the growth curve." The argument goes like this: the rate of growth of the human population is not so steep as it was before; therefore, we may say that it is starting to "level off"; and this looks like "the point of inflection" on the growth curve of small animals in a laboratory experiment. To realize the absurdity of this thinking, it is necessary to say a few words about those animal populations.

When you put healthy fruit flies, or flour beetles, or mice, or flesh flies, in a suitable laboratory cage and give them all the food, water, or bedding they need, they engage in healthy reproduction. The numbers in the cage begin to grow. You put in fresh food and water daily, more than enough for their needs, making every effort to keep them comfortable. The population begins to grow more and more rapidly, geometrically, exponentially, faster and faster and faster. The growth curve by now looks like one of those horror charts of projected growth of the human population from sensational "ecology" literature. And then the rate curve levels off; there is indeed a point of inflection when the population ceases to grow. With beetles and fruit flies the system can be rigged by rationing food or space so that numbers remain stable indefinitely, generation after generation. When you use mice they mostly die out. But all the populations "inflect" and come under "control." It is to this history that we are invited to compare the recent progress of the human population.

The laboratory populations "inflect" because their cages become so crowded that the animals have to struggle for food; or because they no longer have space for some of the vital activities of their niches; or because they blunder into each other and bite by mistake; or because they eat each other's eggs. These troubles interfere with the breeding efforts of the animals. Young are not raised when they get eaten by mistake; eggs are not laid when the mother is starving. The birth rates go down because of privation, and the death rates go up through similar privation. Is this what is happening to the human population? When the population growth of mice in a cage finally stops, one of the things that happens is that mothers eat their babies, definitely making the population "inflect." The absurdity of comparing human history with this is obvious.

Many wild populations of animals, particularly the big ones to which we relate most easily, seem to be constant from year to year, showing that some ancient growth curve must have leveled off in circumstances less drastic than those we engineer in a laboratory cage. But this condition is clearly predictable by niche theory. The animals have a fixed niche and this sets their number. The very last individual for whom there is room is supported and no more individuals can be recruited to the population. Extra individuals are always being produced but the surplus are denied a chance to live. This is the *only* scientific explanation of this kind of population stability that has been found. Competition or predation removes surplus individuals when all the living has been taken up by others. Any other explanation invokes magic.

Are people, all over the world, in their many different densities and styles of living, so pressed by circumstance that no more individuals are being recruited to the population? The idea is obviously and blatantly absurd. It is necessary to make so strong a point of this because various senior physicians, who ought to know better, have been making public statements that the human population is showing signs of being near "its point of inflection."

All contemporary national populations will continue to grow, though at different rates. Only in local places with extreme poverty are populations likely to remain static. These are places where even food is so short that the optimum number of children is less than replacement. Only a very small portion of the humankind is in this desperate extremity. History, therefore, will proceed in its ancient ways, bringing both the good and the bad, with aggressive war definitely on the agenda.

True aggressive war is armed robbery in the name of a political power, be it a city-state, nation or continental empire. It is a war with material gain as its object. The resolve to war comes from ambition and hope. The ecological hypothesis predicts this; the historical record confirms it.

The first requirement of aggression is a rising standard of living. Niches of the ruling classes of the aggressive population have been getting broader, requiring more and more resources for each person. The ruling class will have worked to spread the new standards to poorer sections of the community and there will have been a history of partial success for this effort. More and more of the people will have been living better.

A high standard of living always includes more chance to choose a path in life and is, therefore, seen as a form of freedom. Aggressive armies fight for loot to support a standard of living, but their spokesmen talk of fighting in the cause of liberty. This is not just dishonest propaganda, even when the aggressive armies are led by a tyrant. Alexander, Napoleon and Hitler were despots, but their armies and people all looked for freer times after they had completed their conquests. The English in America talked of freedom more eloquently than any power in history, but they tolerated a standard of living based on slavery over much of their land, and they were so successful in their aggression as to remove the original inhabitants from the whole subcontinent. The belief that you are fighting for liberty is a second general requirement for a war of aggression.

A rising population is a third requirement. This condition will automatically be met when the standard of living is improving and there is a sense of greater freedom. The breeding strategy then works to reflect the new confidence with ample children. But notice particularly that this is not the rising population of the "population bomb." It is not an overwhelming number of poor or hungry. It is a press of people living tolerably and schooled to the idea of living better yet. The real pressure is on the broad niches of those holding political power.

A fourth requirement is that much effort has already gone into meeting the needs of the new freedoms by means less costly than aggressive war. The potential aggressor must be well skilled in contemporary technology, or at least in the technology of its immediate neighbors. It will have made good progress at expanding its resources by technique in agriculture, industry and government. It will have a strong merchant class, using the opportunities of travel and trade to increase their personal standards of living. And it will have communities of its own people dependent on providing or consuming the goods of trade for their regular employment. In material things, therefore, the aggressor state must already be comparatively wealthy.

The fifth requirement, and an extremely important one, is that there must be a suitable victim. The ideal victim is a society that is technologically backward by the standards of the aggressor. It will thus have land and resources from which the aggressors know that they can extract a higher standard of living, possibly for more people than the victim did. The barbarian peoples whom Rome attacked fitted this requirement perfectly, and so did the Amerinds and Australians attacked by the British. Few aggressors have victims so suitable as these but always the victim will have something the aggressor can use. The victim must also be in a relatively poor position to prevent the aggression.

The most common object of aggression is land to settle, to administer or to mine, because more land can always be used to yield more broad niches, even if you do little more than govern it. Sometimes the object is freedom to trade with third parties, and to gain more niche space in that way. But the victim holding the land or preventing trade must be, or appear to be, equipped with weapons inferior to those of the aggressor.

All aggressions are attempted from positions of apparent military superiority. This sixth requirement means that the aggressor usually has, not just a large army, but soldiers with superior technique. In the preparation of an aggression, it is vital that the attacking armies appear to have adequate power to do the job. And in all successful aggressions with lasting results, this requirement has in fact meant that the attacking army has weapons or tactics which are clearly superior to those of the victim and which the victim cannot copy. It is the lesson of history that success in war goes to superior skill and weapons, not to numbers, nor to nebulous qualities like "courage" or "leadership," which all peoples possess equally.

There have been many aggressions in which the apparent military superiority of the attacker turned out to be illusory. Sometimes the aggressor was simply mistaken, falling into the ancient error of thinking that size of an army is a measure of its power. The Persians made this error when they attacked Greece and found that they had no way of beating down a phalanx of hoplites. More often, as in all the aggressions within Europe in the last four hundred years, it was found that the winning technique could be copied by the victim, and the aggressor was battered to a bloody stalemate.

Aggression never comes from a poor country against a rich country, except in very special circumstances. It can happen that a nation appears poor by some standards of measurement, but is wealthy by the test of its own history. Such a nation may be in a position to invent superior weaponry and tactics, when it might be able to destroy the armies of states wealthier in material things and take their lands. Genghis Khan's army launched an aggression of this kind, and so did the Gothic barbarians who destroyed the Roman Empire of the West. These people were poor when their wealth is measured against the luxuries of the civilized empires they overthrew, but they were far from poor by their own standards. They had rising aspirations, desires for liberty, and rising numbers. There was little spread between the lives of their own wealthier and poorer classes. And they equipped themselves with the last word in weapons and submitted to military discipline that turned out to be superior to that of their civilized victims.

It should be obvious that very many of the nations of

the contemporary world are growing in ways that must soon let them fit this profile of a potential aggressor. Standards of life, hopes for liberty, and numbers of the people are all rising together. Many nations show a strong interest in military affairs. Whether they will actually go to war will depend on their finding suitable victims.

The next wars will be fought with the weaponry of the European West. This places a big handicap on aggression, since the warfare of firearms is easy to learn. Some of the battles to come will be decided, therefore, by who can make their own weapons and who has free access to supplies from outside. Victory will go to those with the better industrial base to resupply the armies. This is a novel quality that will have profound effects on the history of the next century. But there are also likely to be effects more novel still from the use of nuclear explosives.

For four hundred years the West has been handicapped in wars with its own kind by the impossibility of achieving a decisive victory before the enemy found a technical counter to the attacking armies. But nuclear weapons make it possible to destroy a victim's power to resist so quickly that resistance may become impossible. It is necessary, therefore, to examine the idea of nuclear war from the standpoint of a potential aggressor, asking the question, "Might it succeed?" If the answer to this question is "Yes," then a nuclear war will surely follow.

In the thirty years since nuclear explosives became available, soldiers and pacifists alike have pondered the chances of war between the United States and the Soviet Union. Their musings depend merely on the fact that these two states keep nuclear arsenals and that they say unkind things about each other. It is also true that each keeps adding to its arsenal for fear of the other. But these facts leave many of the requirements for aggressive war missing.

In both the United States and the Soviet Union, standards of living and numbers of people are both rising, meeting two of the prime requirements for aggressive war. But the numbers are not rising rapidly, particularly in the Soviet Union, and both populations are low compared to the potential niche-spaces that they possess. This is dramatically true when the prime resource is land. The Soviet Union holds one sixth of the land surface of the globe, and the United States, although not endowed on quite this massive scale, is still very comfortably fixed for land. Politicians in both societies are able to convince themselves that they are still improving the standards of life for the mass of their people, that their societies are free, and that their philosophy of life is both comfortable and superior. In spite, therefore, of rising numbers and ambitions there is very little pressure on the broad niches of the classes in power in either country.

Wars of aggression are always popular wars, at least until they turn sour with the prospect of failure. People fight for glory, liberty and loot, or because they see a grim necessity in fighting. There are no needs of Russian or American people that would obviously be met by an attack of one upon the other. There is, in fact, very little that can be used to persuade the people of either that the attack would be worthwhile, even if the attacker did not have to reckon with the certainty of being subjected to nuclear explosives in return.

It is the people with the highest standards of living, those who usually control the government or the organs of decision, who stand to gain most in aggressions.

329

Theirs are the broad niches which will be in shortest supply as the numbers in the middle classes rise. But neither America nor Russia has so developed and used up its own potential niche-space that these classes cannot be looked after. This is still generally true in America and massively so in the Soviet Union, where land, agriculture and industry are still undeveloped by Western standards. And these people of the broad niches have the most to lose in a nuclear exchange.

Neither the United States nor the Soviet Union, therefore, has yet developed to the state where there can be a significant motive for any new aggression, even without the fear of nuclear hurt. Both are already continental empires, with many successful aggressions behind them, and neither has yet absorbed all it has won in past conquests. When the people of each must add into their thoughts of war and peace the knowledge that, however successful a war against the other might be, they would themselves suffer very heavily from nuclear explosives, the motive for an attack tends to vanish.

Nor does an attack by either the Soviet Union or America on the other meet the military requirements for aggressive war. It is true that they are both well supplied with the latest in weapons and that they have schooled themselves in modern military theory. But neither of them has any perceived advantage in technique over the other. Clear technical superiority is an absolutely essential requirement for the starting of an aggression, and there is no chance that either power will gain such an advantage. Putting it differently, neither has the qualities necessary to make it the other's victim.

The ecological hypothesis predicts, therefore, that there can be no war of aggression which involves a clear attack by either the United States or the Soviet Union on the other.

The fears of a U.S.–Soviet war have been fed by an obsession and misreading of recent history. It is argued that the great German or World Wars came about from some dynamic of military power and ideology alone, that peoples of different language and culture somehow had to exercise their military strength against each other without any clear end in view. History gives no support to this hypothesis, for the great aggressions always had acquisition and the movements of people as their aim. In the Second World War these motives were abundantly plain. The two great drives of aggressive armies were those of Germany against the East European and Russian plains, and those of Japan for trade and land outlets in Asia. These were aggressions in the classic mold.

Only in the First World War were the purposes of acquisitive peoples somewhat masked. I have shown, however, how the peoples of each of the continental powers all met the requirements of nations likely to launch aggressive wars. Germany, France and Russia were all fresh from wars of aggressive expansion in Africa or corners of Asia, and were indeed rivals in the plundering from these places. When political fumbling started a war they all joined in with gusto. Germany, by far the strongest and best prepared, then went on, twenty years later, to reveal the true aggressive design with Hitler's war. The First World War was war by accident while an aggression was still being prepared. The Second World War was the aggression finally resolved. Neither struggle provides any prototype for a possible battle between America and Russia. There will be no such battle.

In some circumstances both the Soviet Union and the United States may undertake limited aggressions of their own, because their economies and the hopes of their peoples for improving standards of life are very greatly dependent on a large supply of fossil fuels. They will be tempted to take these supplies by force, and they each have the means to do so. If they fight for oil, it is likely to be in collusion, as allies. But their need for this scarce resource is one that they share with other countries, and one that will be a cause of many fights in the decades to come. It is a need that must be examined before asking the question, If the United States and the Soviet Union will not launch a nuclear aggression which countries will?

The well-being of the European West was built on cheap energy. All previous civilizations used energy that was expensive, human labor supplemented with a little work from animals. Energy is the power to do work. It is necessary to most of the dimensions of a broad, civilized niche.

It was failure to find a source of cheap energy that led to economic stagnation in the later days of the Roman Empire. Romans relied on slaves to make things, carry things, and to do for people of cultivated ways those services that make cultivated living possible. This made certain that very many of the people, the slave classes, would always be poor. But an even more serious consequence was that the very high cost of energy meant that the productivity of industry was always very low. Roman businesses could not get ahead; they could not easily make large surpluses; they failed to generate capital. And a poor business income meant a low tax base, a government short of funds, stagnation in the armies,

332

and eventual collapse. Any civilization poor in energy cannot meet the costs of elaborate government and supply needed by crowding numbers.

Even before the industrial revolution, the European West began with a technology base which was better than that of any previous civilization. Then it found the Americas to take its surplus people and let the numbers grow without impossible strains on the costs of government. And then, after two centuries of growth and conquest without a fossil-fuel economy, it found how to use coal and oil to do the work that had been done in other civilizations by slaves. The coal and oil lay on the ground, loot to be had for the cost of picking it up. It was this loot of fossil energy that let the West come within measurable distance of abolishing poverty, despite their rapidly rising numbers. They could generate capital, give opportunities for trade to more people, carry people in and out of cities to use resources of space in turn, build them houses, free them from brute labor and give them time to experiment with their powers to learn. A very large portion of the people have become wealthy in the sense that they have had the chance to learn to live in a broad niche. It is true that pressures of rapidly rising numbers have tended to maintain a subculture of poverty in even the wealthiest cities, but energy has been so cheap that new ways of living could be invented, for a time, as fast as people were bred to fill the new niche-spaces. Yet it all has depended on a very large flux of very cheap energy.

The rise of the West also depended on cheap food. At first the cheapness came from the new agriculture of novel crops and crop rotations, the farming from which the cities of Renaissance Europe and Tudor England were fed. Then came the vast glut of cheap food from

America, that glut which forced the English government to repeal the corn laws and destroy its own farming industry. The English countryside became depopulated despite the massive growth of the British population. Even in America itself a similar thing happened as large areas of New England, once farmed, were given back to the wilderness in the face of competition from prairie wheat and corn. A historian of the future looking at the record of either old or new England from this period could make the same error of historians of the later Roman Empire who imagine that the population was falling.

The next cause of cheapness in food came from applying the new cheap energy to agriculture. Tractors, harvesting and planting machines and, above all, chemical fertilizers lowered the costs of growing food even as they increased the total supply. The cheapness of food from this episode, now ending, was entirely dependent on the cheapness of the very large fluxes of energy used.

There then came yet one further push to cheap food. This was the development of crops such as hybrid corn, a new agriculture that goes by the name of the "green revolution" in the contemporary press. This agriculture is completely and inextricably dependent on a large flux of cheap energy. The ecological engineering that went into making the new varieties is elegant, but the plants are made to rely on our supplies of cheap energy in order to grow at all. An understanding of this dependence of crops on fuel energy is vital to understanding our future.

The total energy that all our crops can trap from the sun is set in ways that we have not been able to alter. Most likely the actual limit is set by access of the plant to carbon in the air, for it cannot make sugar faster than it can get carbon. All crops and wild plants accept this limit alike, and we have not been able to increase this primary production of plants by one iota. What farmers have done is to breed varieties of plant that put down more of their store of sugar into parts that people like to eat. We measure the productivity of a wheat crop by the weight of grain, not the weight of roots, stems and leaves. Cultivated wheat puts much of its energy reserve of sugar into grain whereas its wild ancestor used most of the reserve to maintain healthy roots and stems in the rough and tumble of wild life, but both kinds of wheat had the same sugar to start with.

With the new varieties of the green revolution we have pushed this process one step further. We have taken over many of the functions that a wild plant had to do for itself, and have done it for the plant ourselves, in factories. We do not let the plant hunt out scarce minerals with its roots, we give it superabundant supplies of fertilizer so that it does not have to work for its nutrients. We take away a plant's ability to protect itself against disease and pests, because the plant used to spend part of the energy reserves of its grain to do the job itself. Instead we protect the plant with chemicals. In other words we keep alive, with fertilizer and chemicals, a plant that would have had no chance of hacking it alone, and the energy that its ancestor would have spent in fighting its own battles is then freed for the plant to make more grain. This extra grain, therefore, is entirely dependent on the cheap fuels supplied to our chemical industries; indeed, in a real sense the energy of this extra grain *is* some of the energy from the chemical industry. We are actually eating fossil fuel. And this fuel is soon going to be expensive almost beyond our present understanding.

335

Western society has been built on the treasure hoard of fossil fuel lying loose at the surface of the earth. It is as if we have been living on the loot of some vast and undetected robbery. But the loot is far gone. The oil may be half used, or more. There is still coal, but the best, or at least the most easily reached, is gone. We have bred very large populations to use this cheap fuel so that our use is now at a rate which means that the remainder must be spent far more quickly than what we have used already. And now the rest of the world wants to use fuel as we have done. We must share the swag— what there is left of it.

This means that energy will soon be expensive whereas once it was cheap. It is not that we will run out of energy; it is rather that we will run out of *cheap* energy. Indeed, we already have, though present (1980) prices are still absurdly low by the standards of what will be the norms ten years from now. Oil, and then coal, will soon be so expensive that nuclear reactors will seem economical to run. We can then pursue research into whatever esoteric methods of energy production we like. There will always be energy, but at a very high price. Never again will energy be cheap, plentiful and easy to extract. This is a fact with profound implications for the politics of nations.

Cheap food too has gone forever. The good parts of the earth are all farmed, and the yield does not quite keep up with the demands of the increasing numbers of people. The crops of the green revolution will be extremely expensive to produce as energy prices rise, probably, in fact, too expensive for poorer countries to use them at all. To the extent that these new crops are abandoned, food production will actually fall, requiring that prices go up in response to the increasing imbalance of demand and supply. Demand too will grow as our numbers continue to grow. In the productive agriculture of the West, farmers will have to start economizing in the use of tractors and fertilizer, as their energy costs climb. They will find themselves using more labor, both human and animal. Their yields need not fall, but the price must go up.

We are, therefore, moving into a time when both energy and food will be dear. Many patterns of civilized life are about to change as a result. The spreads of cities will be different, the countryside will be repopulated, there will be quite different patterns of work and play. It may not be something to fear; it may be rather an opportunity, like all change, for the most adventurous to welcome. Perhaps we can dismantle city governments, break monopolies of power, live country lives when we want to, and work in small industries for brave entrepreneurs instead of serving some giant corporation. Change is always good for the brighter spirits, and the high cost of fuel and food make drastic change inevitable. But the new patterns must certainly offer new temptations and straits which might drive nations to battle, even to nuclear battle.

For there to be a nuclear aggression there must be motive, opportunity and means. It is the large advance in the means for an aggressive war represented by nuclear explosives that must, in some circumstances, make the prospect seem profitable. We are used to thinking about how dreadful nuclear war would be, and with cause. Yet, if we are to understand the circumstances in which this dreadful war might actually happen, we must also look at the theoretical advantages that might come to a nuclear aggressor.

Since the twin inventions of firearms and printing brought in technological war and literate soldiers, it has never been possible for an expanding nation to keep an advantage in military technique long enough to secure a conquest. In modern times only the Germans under Hitler have come close to doing this, yet even they could not secure their new frontiers before their own blitzkrieg of tanks and aircraft was turned against them to crush them back. But the invention of nuclear bombs, delivered from space by ballistic or guided missiles, may have changed this. Not only can a country's will to resist be destroyed, but the country also can be depopulated, albeit at the cost of leaving radioactive debris behind. Swift victory, together with conquered land with few people left in it; this is the classic object of the most successful aggressive wars. Removing people from land is an ugly thought, but aggressive war tends to be ugly.

This new opportunity for aggression given by nuclear explosives comes mainly from the swiftness with which a destructive attack can be completed. Any attacker in the decades of the future must know that its potential victim could build nuclear weapons too. Anybody can make a nuclear bomb. The ways are known, all countries are collecting the necessary Ph.D. physicists, it just needs a little money and time—relatively less of both money and time as the years go on. So not only can any attacker make a bomb, but any potential victim can make one too. But the victim may not have the time. It is this fact that has completely rewritten the rules which have guided wars in Europe these last four hundred years.

There is a definite lead time in making nuclear weapons, which means that the attack can be made and the victim destroyed (literally and horribly) before there is time to prepare a counter. And this is even more true of missile-delivery systems than it is of the bomb itself. Indeed, although it is true that any country can make a bomb, it may be a long time before every country can make the necessary rocket.

The ecological hypothesis predicts that an aggression will always be by the wealthy against the poor, and by the power with the latest weapons against a power with inferior military technique. Aggression by a nuclear force fits these requirements. Nuclear attack, therefore, must give new possibilities for aggressive war. The means are there; the victims are true victims, in that they can be beaten down with little chance to resist; and there is the theoretical possibility of taking over land that has lost most of its people. Yet we must ask whether a radioactive desert is worth taking over. The answer is almost certainly yes.

You ought to be able to use much of a country attacked with nuclear explosives within twenty years or so of the attack. I take this figure merely from the personal experience of working round the Enewetak atoll twenty years after numerous "atomic" and "hydrogen" weapons were exploded there. That small atoll had at least as much attention as would a major enemy city in an all-out nuclear war, probably more attention. But twenty years later the blasted islets are covered with trees and bushes, birds nesting in the trees and insects humming in the sun. We have just sent the people back home to live on Enewetak.

If, in a war, the attacker aimed at the cities, it should be possible to use the countryside after a few months, when the worst fall-out dusts have decayed. (Yes, I know it is ugly, but it is going to be thought out by somebody sometime). Twenty years later you could pick up the debris of the cities, in the unlikely event you wanted them.

There is no doubt that you could use quite a spread of the conquered country very soon after the attack, if you were careful in the attack itself. This would be particularly true if, for instance, one of your prime aims was a mineral resource like oil, coal or metal ore. These need not be spoiled by radiation effects at all because hidden underground. And if you had just taken them by a blatant and savage nuclear aggression, you might well find that others would not be eager to dispute your ownership of them. Taking fuel or raw materials in this way could be a profitable undertaking at once. You could delay using the land for food and settlement until your fifth five-year plan—not really a long way off.

Nuclear attack, therefore, has given new possibilities to the ancient history of aggressive war. But, if the possibilities are new, so must be the revulsion. This revulsion is so strong that a nuclear attack may be expected only after a long history of trouble and hurt has made a policy of nuclear war, if not exactly popular, at least tolerable. That trouble and hurt may be coming. But before looking for future signs of trouble it is convenient to examine in some dispassionate manner what kind of nation might find itself drawn to the dire necessity of a nuclear attack. We must ask ourselves what sort of country is rich, free, ambitious, literate, skilled in trade and commerce, but dependent on the living space of other lands for the wealth and freedom of a large populace, if we want to find a likely nuclear aggressor. The answer, of course, is one of the great island or isolated powers, like Athens, Carthage, Venice, Japan or England. These are the states that have gone to war for trade, colonies, and access to markets since the start of written history. They have had to. Rising numbers and rising freedom on their little patches of land always sent their brighter spirits into trade; and the food and goods of trade made the numbers grow. Soon they traded for their very lives. Fighting for their lives came later. And now the commerce of all industrious islands presses ever more strongly on high technology for the goods of trade. This means that the trading peoples of future wealthy islands will always have the skills needed to make nuclear explosives and rockets if they think they need to.

All free and wealthy island nations, therefore, are likely to meet many of the conditions predicted by the ecological hypothesis to be necessary for aggressive war. They have freedom and wealth, but only tenuous access to resources needed to maintain that freedom. They are able to make whatever weapons are necessary. And they may find a likely victim in a supplier continent that comes to deny supplies. Yet an aggression must be prepared with weaponry decisively superior to that of the victim. Island nations may find that their island base lets them use nuclear weapons in special ways to gain this decisive edge.

True island powers, like England, as opposed to those on isolated bits of continental real estate like Carthage or Athens, do not usually occupy just one island but an archipelago. They have little land scattered through much sea, an arrangement that should give very special advantages in nuclear war. The island people can disperse the launching platforms of their offensive missiles among their remoter islets, keeping their provocative missile silos well away from their centers of population. They can even place their missiles on the sea floor. Better yet, since the island people are of necessity maritime and seafaring people, they can mount their missiles in submarines. And for this an archipelago offers possibilities that have been scarcely explored.

Still the best-hidden of contemporary nuclear missiles are those cruising the world oceans in the *Polaris* boats and the like. They tend to have sailed from ports in the larger land masses, for they have been built by the continental powers. But the owners of an archipelago might well think differently. They can make smaller submarines for use in their shallow seas, close to home, but scattered, hiding their sonar echo against the shallow bottom, defended from enemy fleets by their own island-based killer submarines.

Those who argue about making American missiles impregnable now speculate about turning the whole vast country into a Swiss cheese of empty missile silos while they drag the missiles from one silo to another in an endless game of decoy. Perhaps it is because I am island born and bred that this seems so ludicrously funny, because an island nation can achieve the same result by sailing its missiles through and under its shallow seas in boats. Indeed, it achieves a much better effect, for the "silos" are infinitely shifting and there need be no dummies as decoy at all. And the area of shallow seas in its archipelago may be quite comparable to the vulnerable continental mass on which the present great powers live.

As a site for a nuclear offense, therefore, an island base is probably ideal. But there are advantages that may be as great from the point of view of defense also. This is because the island centers of population might well be defendable against missile attack. The target to be protected is concentrated, and the means of defense are, like the island's own attack, missiles, scattered through the islets and seas of the archipelago. Antimissile missiles are never likely to make much sense for a continent, which has so much to defend, but they might well be practicable for an archipelago.

It may be argued that the crowded peoples of an island are not so much in a good position for nuclear defense as they are prime targets. But, for the sort of nuclear aggression likely to be planned by an island nation, this is not so. Defense against a superpower may be vanishingly difficult, as it is for the superpower itself, but an island aggressor would be arming against a softer enemy than a superpower. Nuclear defense has to be organized with the power of the victim in mind, probably some useful piece of real estate that is not very well run at the moment—say, the future counterpart of Manchuria at the time of the Japanese attack, or the Poland of cavalry soldiers when Hitler's Germans wanted their land. With the proper technical discrepancy between the aggressor and victim, an island nation might be in a very good position to make not only its offense believable, but its defense as well.

Probably what a would-be island aggressor would have most to fear would be that the intended victim would find a superpower for a friend. Even then, if the island planned carefully, the thing might still be done. Once the submarines were deployed throughout the archipelago and the antimissile missiles were in place, it would take a very noble-minded superpower friend to want to do anything real to stop the aggression. It just might get hurt minding someone else's business.

A wealthy, free, industrious and numerous island people can very easily be nudged into a position in which it fits all the requirements for starting a war of aggression with nuclear explosives. It would, of course, only do so grim a deed in a world so changed that the act may be both thinkable and even "necessary." But the world will change.

Type specimens of warrior island states are England and Japan. Both have, at times, found the aspirations of their peoples frustrated because the home islands could not be made to yield more resources quickly enough, and both have always responded by aggression against their neighbors. England built the largest empire in the history of the world, and Japan only a generation ago was tempted to make a vain assault with inadequate weapons against an immense continental power. We look now on both of these nations as being rather peaceable, sensible states, offering moderate counsel in the world's debates. It is rather startling to remember that probably one third of the living males in each of these two countries has seen active service as a soldier in armies as tenacious as any in history.

Other island powers will rise over the next century. They too will be forced by niche and breeding strategy to that same sequence of success; liberty, rising numbers, trade, more liberty, and rising numbers again. This is a progression that leads to the need to fight; indeed, to attack. Yet all this, though likely, may be a hundred years off. Of more immediate interest to the West is the course of life lived in the rich continents already sheltered by nuclear arsenals.

There are three technically modern, more or less self-supporting, present-day empires: the United States, Europe and the Soviet Union. Each of these places is properly to be looked at as a continental empire, put together in the last four hundred years from scattered arrays of small and very different states. Each union is

344

wealthy, literate, and with high ambitions for the lives of the people.

It is more usual to think of the differences among these three powers than of what they have in common. The empire across Asia was built by Romanov kings, by aggression using European weapons mastered in the early days after Swedish armies were stopped from preventing a Russian imperial future altogether. The Romanovs put together a classic empire of conquest of less-developed peoples, and then lost control to a new system of government. But the empire remained. The European Union, still being cemented, was built out of the seesaw battles between nation-states of equal competence and technique in war, very much in the way that ancient Greece lurched toward eventual union under Macedon. And the United States conquered a stone-age continent and then arranged for it to be re-peopled with immigrants from all the cultures of Europe.

But the similarities are much greater than these differences. Each empire has a resource base that cannot be made much larger, because each has already taken the prime land within reach. Each is essentially impregnable to powers outside the privileged three themselves; which means that each can expect to keep the resources it now has for as long as it maintains its technical and military strength. Thus, although the three empires are not likely to grow larger, they will not shrink either. Their condition has echoes of Rome in the good Antonine years, when the Empire lived in peace behind the dike of legions. But the modern dike is nuclear.

Perhaps it is necessary to remark on the fact that Europe is still in the little league of nuclear armaments.

This is a passing fact of contemporary history, made possible because Europe lags, if only for the time being, behind America in weapons. In terms of actual technical wealth (and skill to be turned into nuclear defenses if need be) Europe is now probably the strongest of the three.

We are still tempted to play with scary dreams in which American weapons are kept aside long enough for a Soviet attack to "annihilate" Europe, but nightmares like this do not take into account any reasons that could make such an attack plausible. Even without the hideous risk to itself, there is no obvious gain that could come to the Soviet Union from such behavior. These arguments are the same that were considered earlier in rejecting the possibility of a war between the United States and the Soviet Union. Europe is already a pointless target for nuclear attack, and we may be sure that the coming decades will find it more obviously and dangerously unattackable as well.

Each of these three empires, while owning essentially all the land it will ever hold, is well filled with people grown accustomed to high standards of living. Moreover, the people have grown used to the idea that the standard of living will continue to improve. All have learned to use energy from fossil fuels to multiply the benefit from each natural resource many times, providing many broad-niche spaces out of the land they own. They have each driven poverty down so that in each empire the number of people who must be poor is proportionately smaller than in any society since those early times of hunting and gathering, when there were no poor. And these fortunate people have the command of such weapons that none can attack them. They are at a plateau of prosperity, like earlier empires in their primes, but on a grander scale.

And yet the process of historical change can be seen at work in these societies too. The breeding strategies have not changed in any of them, and every Russian, American and European couple continues to raise the number of children they think they can afford. Every generation sees more people for whom an affluent life must be provided. Some of the effects of this are now obvious enough, even in America, and they have been apparent in parts of Europe for longer. Access to the wilder places and the country becomes restricted and rationed by price. In Europe the mass of the people have long been denied the use of wilderness or country-side by patterns of "ownership" that make "no trespass-ing" a common sign of law. Americans are still happily ignorant of laws against trespass, yet they find fewer and fewer places where they can go without checking with some official first. Americans must reserve time to climb a mountain, file travel plans if they walk in the Sierras, get permission before they wander in an Alas-kan wild place. We can no longer do as we please be-cause so many people want the land that they cannot all use it at the same time. So the land is rationed—though various euphemisms are used for the offensive socialist word "ration."

When we use wild land we allow ourselves those un-changing dimensions of our ancient ice-age niche that were measured in space. It is easy to see that rising numbers would soon press upon these dimensions of niche, and yet we know we have lost something real when we let it happen. To hunt freely, travel where you will, light a fire and spread your sleeping roll; our grandfather's generation could do more of those things than we can. One of the freedoms, albeit perhaps a lesser one, has clearly gone before the press of numbers.

But other freedoms, more precious than these, have

347

been under attack also, even under the most libertarian of constitutions. City, suburban and business life is set about with regulations—irritating, pettifogging, bureaucratic restrictions. We blame governments for being too big and remote but, whether the mood of the electorate swings to the left or the right, nothing much seems to change. Yet it is not some error of government that causes this restriction, it is the gentle jostlings of the people. It is a result of people-pressure. The irksome mounting of petty restrictions, which president and prime minister alike have not been able to stop, is the fruit of expansion when the numbers of people are only a little fewer than the number of opportunities there are to let them live in a reasonable way. The people must be rationed to niche-spaces, and bureaucratic restrictions are the ration cards.

There must now be fear that the press of restriction will increase, possibly rapidly, because we are about to lose our large flux of cheap energy and cheap food. Almost inescapably, lack of cheap energy will mean lack of cheap capital, which will lead to a progressive shortage of new opportunities for living well. Since the numbers of people must be expected to continue slowly to rise, then the progressive loss of freedom that we already experience must accelerate.

It is easy to compare the condition of our three wealthy empires with others in the same stage in a historical cycle in earlier times, and to feel depressed. Always numbers pushed against resources and capital so that freedom was curtailed, and we can see clear enough echoes of this process in our own affairs. The older societies always developed very oppressive social systems when the rising numbers could be accommodated in no other way; the mass was compressed so that the few might live well. Likewise we find ourselves beset by the big government which is a part of this process. We may not yet be ranked into various degrees among our social betters, but we are becoming increasingly supplicant before officials who say what we may and may not do. If we do not find ourselves ranked more steeply by social caste, it is because we have earlier gone so far in removing poor, narrow and low-caste lives from our societies entirely. Perhaps universal literacy will keep all of us, Europeans, Soviets, and Americans alike, from drifting back into systems of caste, but then each society will find other ways of keeping people in their places. Probably this means state socialism with its idea of equal shares of what little there is, backed up by the sanction of law. Our choice, therefore, will be rationing by caste and wealth to yield unequal shares in great variety or rationing by the apparatus of a socialist state with its inevitable uniformity.

Liberty, in the Jeffersonian sense, cannot survive a continual packing-in of people. If our numbers continue to rise on a resource base that expands but little, the future inevitably holds ever greater restrictions on individual freedom. Our descendants will not be able to live as we live and our free American and European ways of doing things will seem like poems of the past. Liberty will fall progressively as the numbers rise, and obedient compliance with the majority Will must take the place of individual initiative. Perhaps some politician cleverer than the rest will arrange this necessary peaceful compliance and call it "free."

Yet the future may not be so dull as this, for we have big novelties working for us. We have knowledge far transcending anything known by earlier crowding empires, knowledge both of how to make things and for

349

understanding our own human habits. We gleaned this knowledge during the years of cheap energy, when the immense glut of resources freed unprecedented numbers of people to tinker with machines, to inquire into the esoteric or useless, to think, to research. We have not wasted the fossil-fuel century because we have converted enough of the glut of energy into the highest of formally ordered forms, "information." We shall go on moving peoples and planning new ways of life out of this knowledge won from energy that was almost free, and the knowledge cannot be taken from us easily. No old civilization had this knowledge. As long as we are busy using it, as we certainly are going to need to be, there will always be a sporting chance of keeping the socialist bureaucrat off our backs. But there may be more salvation still in what we know of the workings of people themselves.

We know enough of our biology now to realize how unique we are. We, alone of all species, learn most of the habits that fashion our lives. Learning is the uniqueness, the essential humanity, of our kind. Very little we do still comes from animal compulsion, not even raising children. All learned tasks can be undertaken equally well, or equally badly, by people of either sex. Furthermore, the learned part of our niche is so much the larger part that there is no biological basis for claiming that principal functions of individuals or individual sexes are set by nature.

Part of this has long been obvious, for it is now old-fashioned to think that males have to fight or dominate others in order to be happy or even natural. By the same token it is a gross travesty of humanity to suggest that the natural role of a woman is raising children. It is a woman's good fortune that she can raise children; that is an animal prize she holds. But all her essential humane qualities have nothing to do with animal urges, but are learned, people things. Roles that people learn are more human than the roles passed to them in passivity, like the roles of a worm.

Probably even churches will come to recognise that women are humans and that in this human species no sex is set aside from daily life for the purpose of raising children. In some countries this change is already far advanced. And a consequence may well be that our breeding strategy will yield even smaller families for the moderately affluent than they have now. Our breeding strategy can remain the same, leaving every couple to decide the number of children they think they can afford, but the more nonchild fulfillments in the lives of a married couple, the smaller may be the number they choose as optimum. If this should happen, if welcoming women as social equals of men does lower the size of Western families, then the remorseless press of more people in our wealthy empires may slide away. Resources and opportunity that our knowledge can glean, even in a world of dear energy, can then creep ahead of demand and a very free life might result. We could even think of doing something constructive for people outside our nuclear dikes.

So the future in the three modern empires may not be so bleak as simple historical precedents might suggest. This is a good time to be living, for ours are the generations with accumulated knowledge and who yet see the end of the easy times with their swag of free energy. Change, the friend of the clever and the innovative, is close upon us. There are going to be some good and interesting things to do.

Safety from war, peril to liberty from rising numbers,

hope that new roles for women will hold down the population and let us all live in freedom; these three decide our future for the next few centuries. More detailed speculations are unsafe, but there is one provocative possibility tempting to write down as a hostage to fortune. Perhaps a time will come, say fifty or a hundred years from now, when there is more real liberty of the Jeffersonian kind in the Soviet Union than there is in the West. I base this quixotic speculation purely on the niche-theory argument that the resources in land commanded by the Soviet Union are so vastly greater in proportion to the numbers of their people than the resources of the other empires. The Russians are training their people to live in literate and expansive ways. Although the Soviet state still finds it hard to let women be both mothers and workers, the Russians have gone further in accepting women in society than we have, and this has already paid them well in a more modest growth of population. With much learning, moderate numbers, good weapons to protect them, and a very large land, little more than an excess of policemen prevents Soviet Russia from being in an enviable free state. And policemen come and go.

But while people in the three modern empires work out their futures behind their nuclear dikes, lengthy and bitter wars will inevitably be fought in the struggling world outside. In all three of the tropical continents the rapid crowding of people is already bringing its inevitable frustrations. These are of a scale never endured by any civilization before, because people are being made literate and taught the desires and aspirations of wealthier states, even as their poverty grinds on. Crowded masses are not only affronted by the comfortable living of their own rich, like the crowded masses of ancient empires, but they must constantly be told of people in contemporary lands who need not endure such misery. And, along with the knowledge of other peoples' ample lives, comes the knowledge of other peoples' weapons.

Automatic firearms, bombs and antitank grenades make it possible to arm a modern mass so well that the concept of the citizen soldier, as it was understood by the ancient Greek city-states, becomes once more plausible and real. A gifted general in one of the crowded tropical countries may be better furnished for soldiers than even was Napoleon. Napoleonic armies had wild enthusiasm, but only muskets to shoot with. The coming armies of the equator will have weapons so deadly, yet so simple to use, that every able-bodied person can soon become worth having as a soldier. We must expect that generals in these states will be able to call upon not just huddled masses, as did, say, the kings of ancient Persia, but on deadly fighting men, as did the Greek leaders who called out their armored spearmen.

Africa holds the greatest possibilities for the aspiring general. It is now divided into many small states all of which are learning one of only two common languages, English or French. These languages may fade, but they must leave a heritage of grouping and blocks with their passing. Meanwhile, each small country was given Western hopes by fading colonial powers, when the countries were in no economic condition for politicians to be able to meet such hopes. Elected power never lasts if it cannot fulfill a promise, and the new leaders of Africa had been pushed by circumstance into promising the impossible. It is inevitable that African countries should usually come under military rule.

And the numbers of the people must rise in every

African country. This has little to do with Western medicine, nor is it to be blamed on African tribal systems unsuited to modern techniques of living. The rise in numbers is the expected response of the human breeding strategy to times of hope. African families are large because the ways of the European West have reached Africa and are producing more food, better houses, migration to the cities, and an atmosphere of hopeful change. The breeding strategy is expected to yield large families in these circumstances, just as it did for Europe and America when they came that way themselves. But the rate at which these new people arrive makes their aspirations hard to meet, as it always has done. As in the European past, these times of hope and rising number make African people ready to follow soldiers who will promise a living through fighting.

In a continent where the numbers and habits are changing so very fast, there must always be squabbles ready for soldiers to settle. Local battles will be won and lost. Local tyrants will rise and fall, tyrants whom we can best understand if we measure them by the appropriate histories of our Western past. Idi Amin had vile habits and a private torture chamber, so the world seems a better place in which to live now that he has fallen. But Peter Romanov too had vile habits and a private torture chamber just as foul, yet he appears in the history books as Peter the Great, the victor of Poltava and founder of the modern Russian state.

As local African battles are won, so there will be new and more powerful enemies at the border for an African general to fight. It may be a larger version of the wars between Greek city-states, when Sparta learned her trade, or like the rapid passage of victory vaulting over victory that brought Genghis Kahn to power. The details and the durations of African wars will depend as elsewhere, on the relative disciplines of local cultures and religions, and on the chance of genius that throws up the captains. But that there will be battles between African nations as they build their African continent in a new image is as certain as anything in history. For each country there must come times when wealth, hopes, ambitions and numbers all rise together. It then needs only access to high-quality weapons for an aggression to be an attractive undertaking.

No doubt African wars will be complicated by interference from outside powers, both peripheral and distant. People in the three great empires will send both weapons and advisers to one side or the other in almost all fights because this gives employment to their own people. Backing a side in an African war will serve the cause of a limited aggression in this way, giving outlets to manufactures and people who go a'soldiering from Europe, Soviet Russia and America, as conquest and trade always do. Recently Russian weapons in Cuban hands have been to the fore in this process, but this is just a beginning. Perhaps moral rectitude will grasp the governments of all three empires simultaneously so that they leave Africans to their necessary fighting without "help," but history offers little precedent for such self-denial.

People round the fringe of Africa may have more decisive impacts than meddlers from the empires because they often have selfish interests in the power or weakness of black Africa. The Persian and Arab states north of the Sahara, inheritors of long cultural traditions held intact since before the days of Rome, show signs of becoming wealthy, literate and free. They will surely use money and armies to decide African fights in ways suited to their polity. Populous Egypt still brokes the power of the Nile Valley after five thousand years and is rapidly mastering the skills of Western industry; she cannot look at African war without thinking of Egyptian interest. And neither Israel nor the Dutch descendants at the Cape will find it easy to remain aloof from African fights, however much they may wish to keep clear.

Large wars in Africa are certain; because the necessary mix of many governments, rising wealth, rising hopes and rising numbers will be present for many years to come. Interference from outside powers is also certain, inevitably making the wars more bitter and prolonged. But the African fights will not be nuclear, they will be fought by infantry using automatic weapons. In the more prolonged wars this will become increasingly evident as the high cost of fighting vehicles, whether flying or crawling, makes their replacement ruinously expensive. Only when one of the three imperial powers lets itself be dragged deeply in are fighting vehicles likely to decide wars, as opposed to fleeting battles. But keeping control of any part of Africa in the service of one of the three empires will be virtually impossible in the face of local indignation supplied with automatic handguns. African armies will take African wars to African solutions with weapons they can make themselves, yet the fighting may be less protracted than the wars which settled the shape of Europe as Africans learn skills worked out in other countries. Even if wisdom triumphs so that African unions are made with the minimum of war, it is probably true to say that any young man of the next century who dreams of writing his name in the dubious honor roll of the great captains had best be black.

In tropical Asia the wars of Union were fought centuries ago. They left people living in crowded empires, with elaborate caste structures, permanent poverty, and ancient religions well-suited to life at high densities. And to this has been added both the aspirations and weapons of the European West. Asians have improved material living as well, by bringing the Western trick of multiplying the niche-space with a flux of cheap energy. Ways have been found to escape from the ancient poverty for a time, and families have grown larger with the new hopes. This has brought anew that fatal combination of rising wealth, rising numbers, and access to effective weapons—the necessary precursors to aggressive war. In Southeast Asian countries, the result has been civil war, the well-worn and predictable path of the American and French revolutions as aspiring people armed themselves to make new governments. Then, again as in the Western models, revolution has spilled over into aggression against neighbors. Even as I write, people such as those we call the "boat people" are being driven out, whole populations at a time in the ancient pattern of aggression following successful revolution. For these are all lands very full of people, and victors still prefer uncluttered land.

Fortunately, the ancient empires of India and China still stand and grow stronger. They even follow the wise policy of trying to repair the damage done to their people's liberties by excessive population. And both empires have taken what may be the wisest step of all in that they have built themselves nuclear explosives. It must be humane to give a large people a breathing spell from petty attack while they plan the future in decency. Let us not grudge them nuclear weapons; those large and learned nations have real need of them.

357

And in the third tropical land mass, South America, there are chances for warfare also, perhaps on the pattern of eighteenth-century Europe. These are countries of rising literacy, rising ambitions, rising civilization and rising numbers. They have armies planned on the principle of the expanding torrent. The lands are of different sizes and differently endowed with natural strength, making it feasible enough to pair the countries off into potential aggressors and victims. In South America, therefore, all the conditions stated by the ecological hypothesis to be necessary for aggressive war are met. And already most South American countries tend to be ruled by soldiers.

It seems a safe prediction that there will be modern wars of dash and movement in South America; yet it also seems safe to say that the map will be little changed by the fighting. It will be as it was in Europe, with a victim and a victim's friends being able to collect the latest weapons of war in time to put back any conquest in the end. Perhaps the final history of warfare will include the Spanish or Portuguese name of a South American great-captain, who scored extraordinary victories for ten or twenty years before being crushed in some Amazonian or Andean Waterloo. But a South America des Patries is likely to be the real outcome.

The future that is coming will be diverse because the world is a very large place, with much variety in its people. Strange combinations and pressures will come together. The numbers of the people will rise for a long time in most countries. Nations of growing wealth will war on the weak. Other nations whose liberties seem at stake will fight because they feel cornered. There will be nuclear attacks, probably from a wealthy island against a technically backward supplier of raw materials or food. None of the three superpowers, however, will fight in nuclear wars and there will be no nuclear exchange of the kind imagined in nightmares that pit the United States against the Soviet Union. Nuclear strikes that do come from the smaller powers will have no lasting consequences to countries not directly involved. There will be wars over oil. There will be large economic failures as the years of cheap energy and cheap food end forever. Africans will fight in wars of national union. There will be tyranny and revolution and liberty.

By far the worst prospect we have to face is that the freest of us will lose our liberty from a remorseless and gentle jostling of crowds of people. This dread future is certain if we let our numbers grow even as our economies, bereft of cheap energy, lag behind. Yet even here there is room for optimism and hope. We did put the years of rapid growth to good use, learning to be clever in technical things as no civilization did before. We learned to go on learning. Above all, we no longer have to think of our bodies as mysterious works of magic. We are human, understandable, and very different from other animals. We breed in human and controllable ways. We can change our life styles to let women do more useful things than raise surplus children. Because we can work out what is happening to us, we need fear neither our future nor our fate.

A Note
on Source
and Substance

The ecology on which I base my theory of history is a modern subject only now beginning to emerge from scholarly writings in scientific journals. I have given a general account of the growth of the ideas of modern ecology in *Why Big Fierce Animals Are Rare* (Princeton, N.J.: Princeton University Press, 1978) and in an earlier textbook, *Introduction to Ecology* (New York: Wiley, 1973). Any college-level textbook on evolutionary ecology published since 1970 will contain discussions of niche theory and will supply references to the growing primary literature on this subject. I list a few useful books below. It is necessary to be cautious when choosing something to read on "ecology," because of the many and varied meanings now put on the word. Any public library will have a thick collection of cards filed under "ecology," but it is unlikely that many of the books in the collection will have much to do with niche theory, the evolution of species diversity, or of breeding strategy; instead, they either will be about the environmental crisis, real or imagined, or will even describe curious "back to nature" political philosophies quite foreign to the thinking of a professional ecologist. But even

in the primary ecological literature there is no other
attempt to apply niche theory and breeding strategy to
human history as I have done.

I began to think in these ways in the late 1960s and
first developed a formal theory of history in 1971. For
a university teacher that academic year began with the
most extraordinary arrival of ecology as a popular subject, when great demonstrations were held and much
nonsense spoken about the impending collapse of life
on earth. I announced a course of lectures in the spring
of 1971 at The Ohio State University to examine what
ecological theory truly had to say about the human condition, and soon I was lecturing live to a thousand students at a time in a great crowded auditorium. And
then, almost as soon as my lecture series began, our vast
campus erupted into riot, tear gas, and dangerous confrontation between crowds several thousand strong and
marching lines of armed men. I worked out my ideas of
the human niche in revolution as I wandered through
the tear gas, argued, and looked into the muzzles of
those M.1 carbines in the scenes I describe in the American chapter. Then I took me to my great auditorium
and pursued the logic of my ideas with that remarkable
large class. In a later year I went away to a little mews
house on the edge of Hyde Park in London for a year
with a fellowship from the John Simon Guggenheim
Memorial Foundation, which was when I wrote a first
draft of this book.

The history of which I write is mostly the common
knowledge of Christendom, but my thesis has set me to
examining most particularly the events around the
great aggressions and the battles with which the great
captains recorded the supremacy of new power. I took
my battle scenes from General Fuller and Liddell Hart

as well as many a specialist account. But I have always gone back to more direct accounts when possible. This is easy for the battles of classic antiquity, because English translations of Livy, Polybius, Tacitus, Arrian and the rest are simple to read and cheap to own. Their accounts may not be accurate, but they are free from the imagination of some later commentators. Though I am aware that Creasy is thought to be an "unsound authority," I still recommend his "fifteen decisive battles of the world" for superb and generally accurate battle scenes. The clearest insight into what gave the victory in many a crucial battle is to be found in the work of that most unprofessional historian, Field Marshal Montgomery.

The overview of history's facts which I use borrows much from Toynbee, and part of my historical education was to read my way through his *Study of History*. My interpretation of history's facts is, of course, very different from Toynbee's, but his immense synthesis makes available to students of other disciplines a body of knowledge which is hard to acquire from other sources. I give below a reading list from which the essential facts of climacteric history can be gleaned.

ECOLOGY REFERENCES

COLINVAUX, P. A., *Introduction to Ecology*. New York: Wiley, 1973 (610 pages).

COLINVAUX, P. A., *Why Big Fierce Animals Are Rare*. Princeton, New Jersey: Princeton University Press, 1978 (256 pages).

HUTCHINSON, G. E., *An Introduction to Population Ecology*. New Haven and London: Yale University Press, 1978 (260 pages).

KREBS, C. J., *Ecology: The Experimental Analysis of Distribution and Abundance*. New York: Harper and Row, 1972 (694 pages).

KREBS, C. J., and Davis, N. B., *Behavioural Ecology: An Evolutionary Approach*. Sunderland, Mass.: Sinauer Associates, 1978 (494 pages).

MACARTHUR, R. H., *Geographical Ecology*. New York: Harper and Row, 1972 (269 pages).

McNAUGHTON, S. J., and Wolf, L. L., *General Ecology* (2d edition). New York: Holt, Rinehart and Winston, 1979 (702 pages).

ODUM, E. P., *Fundamentals of Ecology* (3d edition). Philadelphia: W. B. Saunders Company, 1971 (574 pages).

PIANKA, E. R., *Evolutionary Ecology* (2d edition). New York: Harper and Row, 1978 (397 pages).

RICKLEFS, R. E., *Ecology* (2d edition). Newton, Mass.: Chiron Press, 1978 (861 pages).

ROSENWEIG, M. L., *And Replenish the Earth: The Evolution, Consequences, and Prevention of Overpopulation*. New York: Harper and Row, 1974 (304 pages).

SMITH, R. L., *Ecology and Field Biology* (2d edition). New York: Harper and Row, 1974 (850 pages).

SOUTHWICK, C. H., *Ecology and the Quality of Our Environment* (2d edition). Princeton, N.J.: D. Van Nostrand Company, 1976 (426 pages).

HISTORY REFERENCES

Arrian, *The Campaigns of Alexander*, trans. A. De Selincourt. Baltimore: Penguin Books, 1958 (430 pages).

Barnett, C., *The Collapse of British Power*. London: Eyre Methuen, 1972 (643 pages).

Birley, A., *Septimius Severus, The African Emperor*. Garden City, N.Y.: Doubleday and Company, 1972 (398 pages).

Churchill, W. S., *Marlborough: His Life and Times* (4 vols.). London: Harrap, 1933–38. (See particularly Vol. II for the genesis and battle of Blenheim.)

———, *The History of English Speaking Peoples* (4 vols.). New York: Dodd, Mead and Company, 1958–62.

Clausewitz, C. V., *On War*, trans. J. J. Graham (3 vols.). London: Kegan Paul, 1908.

Creasy, E. S., *The Fifteen Decisive Battles of the World*. (Numerous editions, that of 1854 has 639 pages.)

Eburne, R., *A Plain Pathway to Plantations*. Ithaca, N.Y.: Cornell University Press, Folger Library edition, 1962 (154 pages).

Einhard the Frank, *The Life of Charlemagne*, trans. L. Thorpe. Baltimore, Penguin Books, 1969 (83 pages).

Foch, F., *The Principles of War*, trans. H. Belloc. London: Chapman and Hall, 1919 (351 pages).

Fox, R., *Genghis Khan*. London: Bodley Head, 1936 (285 pages).

Fuller, J. F. C. *The Decisive Battles of the Western World* (3 vols.). London: Eyre and Spottiswoode, 1954–56.

———, *The Generalship of Alexander the Great*. London: Eyre and Spottiswoode, 1958 (319 pages).

Gibbon, E., *The History of the Decline and Fall of the Roman Empire* (8 vols.). London: T. Cadell, 1838.

Grant, M., *The Climax of Rome*. London: Cardinal, 1974 (349 pages).

———, *The World of Rome*. New York: World, 1960 (349 pages).

Graves, R., *Good-bye to All That*. London: Jonathan Cape, 1929 (446 pages).

Herodotus, *The Persian Wars*, trans. G. Rawlinson. New York: Modern Library, 1942 (714 pages).

Hitler, A., *Mein Kampf*, trans. J. Murphy. London: Hurst and Blackett, 1939 (568 pages).

Hocart, A. M., *Caste: A Comparative Study*. New York: Russell and Russell, 1968 (157 pages).

Ishimaru, T., *Japan Must Fight Britain*, trans. G. V. Rayment. London: Hurst and Blackett, 1936 (288 pages).

Jones, A. H. M., *A History of Rome Through the Fifth Century* (2 vols.). London: Macmillan and Company, 1968.

Kahn, H., *On Thermonuclear War*. Princeton, N.J.: Princeton University Press, 1960 (651 pages).

Kitson, F., *Low Intensity Operations*. London: Faber and Faber, 1971 (208 pages).

Liddell, H. G., *A History of Rome from the Earliest Times to the Establishment of the Empire*. London: John Murray, 1909 (750 pages).

Liddell Hart, B. H., *A Greater than Napoleon: Scipio Africanus*. Boston: Little, Brown, 1927 (281 pages).

———, *Paris, or the Future of War*. New York: E. P. Dutton, 1925 (86 pages).

———, *The Revolution in Warfare*. New Haven, Conn.: Yale University Press, 1947 (125 pages).

———, *Strategy: The Indirect Approach* (revised edition). London: Faber and Faber, 1967 (430 pages).

Livius, T., *Books 21–30 of Livy's History of Rome from Its Foundation: The War With Hannibal*, trans. A. De Selincourt. Baltimore: Penguin, 1965 (711 pages).

Mao Tse-tung, *Selected Military Writings of Mao Tse-tung*. Peking: Foreign Languages Press, 1968 (410 pages).

Martin, H. D., *The Rise of Chingis Khan and the Conquest of*

North China. Baltimore: Johns Hopkins Press, 1950 (360 pages).

Maurois, A., *A History of France,* trans. H. L. Binsse. New York: Minerva Books, 1968 (598 pages).

Mellenthin, F. W., *Panzer Battles, A Study of the Employment of Armor in the Second World War,* trans. L. C. F. Turner. Norman, Okla.: University of Oklahoma Press, 1956 (383 pages).

Moltke, H. von, *The Franco-German War,* trans. C. Bell and H. W. Fisher (2 vols.). London: James R. Osgood, McIlvaine and Col, 1891.

Montgomery, B. C., *A History of Warfare.* London: Collins, 1968 (584 pages).

Morison, S. E., and Commager, H. S., *The Growth of the American Republic* (2 vols.). New York: Oxford University Press, 1962.

Plato, *The Last Days of Socrates,* trans. H. Tredennick. Baltimore: Penguin Books, 1958 (168 pages).

Polybius, *Polybius: The Histories,* trans. M. Chambers. New York: Washington Square, 1966 (340 pages).

Pratt, F., *The Battles That Changed History.* Garden City, N.Y.: Doubleday, 1956 (348 pages).

Prothero, R. E., *English Farming Past and Present.* London: Longmans, Green, 1912 (504 pages).

Remarque, E. M., *All Quiet on the Western Front,* trans. A. W. Wheen. Boston: Little, Brown, 1929 (291 pages).

Sassoon, S., *Memoirs of a Fox-Hunting Man.* London: Faber & Faber, 1928 (395 pages).

Shirer, W. L., *The Rise and Fall of the Third Reich.* New York: Simon and Schuster, 1960 (1,245 pages).

Story, R., *A History of Modern Japan.* Baltimore: Penguin Books, 1960 (300 pages).

Strachey, J., *The End of Empire.* New York: Random House, 1960 (351 pages).

Suetonius, *The Lives of the Twelve Caesars,* trans. J. Gavorse. New York: Modern Library, 1931 (317 pages).

Tacitus, *The Agricola and the Germania,* trans. H. Mattingly and

S. A. Handford. Baltimore: Penguin Books, 1970 (175 pages).

Thucydides, *The History of the Peloponnesian War,* trans. R. Crawley. London: J. M. Dent and Sons, 1945 (489 pages).

Toynbee, A. J., *A Study of History* (12 vols.). London: Oxford University Press, 1934–1961.

——, *A Study of History,* Abridgement of Volumes I–VI by D. C. Somervell. New York and London: Oxford University Press, 1947 (617 pages).

Varley, H. P., *Imperial Restoration in Medieval Japan.* New York: Columbia University Press, 1971 (222 pages).

Vladimirtsov, B. Y., *The Life of Chingis-Khan.* London: Routledge, 1930 (169 pages).

Index

371

ABOUT THE AUTHOR

PAUL COLINVAUX, who was educated at Cambridge and Duke Universities, spent his postdoctoral years in Belfast and at Yale. Though much of his own research relates to the history of the environment and climate of the ice-age earth—particularly of Arctic Alaska, the old Bering Land Bridge, the Andes of Ecuador and Peru, and the Galapagos Islands, where he has led several expeditions—he also serves as a field assistant to his scientist wife on dives on coral reefs.

Professor Colinvaux, who is a professor at The Ohio State University, has been a Guggenheim Fellow, a James B. Duke Fellow, a NATO Fellow, and a Visiting Professor at the University of Washington. His two previous books are *Introduction to Ecology* and *Why Big Fierce Animals Are Rare.*